Beyond Signals - Exploring Revolutionary Fourier Transform Applications

Edited by Muhammad Bilal Tahir

Published in London, United Kingdom

Beyond Signals - Exploring Revolutionary Fourier Transform Applications
http://dx.doi.org/10.5772/intechopen.1005547
Edited by Muhammad Bilal Tahir

Contributors
Alessandra Canepa, Antonio Martínez Cortizas, Bidyarani Langpoklakpam, Gargi J. Trivedi, Huthaifa Sameeh Alqaralleh, Lithungo K. Murry, Miomir S. Stanković, Olalla López-Costas, Slobodan B. Tričković, Talbakov Farkhodjon Makhmadshoevich

First published in London, United Kingdom, 2025 by IntechOpen
IntechOpen is the global imprint of INTECHOPEN LIMITED, registered in England and Wales, registration number: 11086078, 167-169 Great Portland Street, London, W1W 5PF, United Kingdom

For EU product safety concerns: IN TECH d.o.o., Prolaz Marije Krucifikse Kozulić 3, 51000 Rijeka, Croatia, info@intechopen.com or visit our website at intechopen.com.

British Library Cataloguing-in-Publication Data
A catalogue record for this book is available from the British Library

Beyond Signals - Exploring Revolutionary Fourier Transform Applications
Edited by Muhammad Bilal Tahir
p. cm.
Print ISBN 978-1-83634-242-7
Online ISBN 978-1-83634-241-0
eBook (PDF) ISBN 978-1-83634-243-4

If disposing of this product, please recycle the paper responsibly.

Meet the editor

Dr. Muhammad Bilal Tahir is an esteemed physicist specializing in nanomaterials, sensors, and renewable energy applications. He serves as Associate Dean of the Faculty of Natural and Applied Sciences and Director of the Institute of Physics at Khwaja Fareed University of Engineering and Information Technology (KFUEIT) in Pakistan. With extensive research contributions in photonics and optoelectronics, Dr. Tahir has authored numerous high-impact publications and led international collaborations. His expertise spans Fourier Transform applications in advanced material characterization. A dedicated academic leader, he also holds the position of Provost at KFUEIT. Dr. Tahir actively engages in curriculum development, quality enhancement, and fostering academia-industry linkages, making significant strides in the scientific and educational landscape.

Contents

Preface

Fourier Transform is among the most powerful mathematical tools used in various scientific and engineering disciplines. From its origins in solving heat conduction problems to its modern applications in signal processing, spectroscopy, and quantum mechanics, Fourier analysis has revolutionized the way we understand and interpret complex data. *Beyond Signals – Exploring Revolutionary Fourier Transform Applications* is a comprehensive volume that brings together leading experts to discuss its diverse and groundbreaking applications across multiple domains.

This book explores the theoretical foundations and practical implementations of Fourier Transform techniques, highlighting their role in solving contemporary scientific challenges. It is structured to provide an in-depth understanding of Fourier methods in areas such as material characterization, biomedical imaging, nanotechnology, and computational modeling. Each chapter presents an analysis of cutting-edge developments, demonstrating how the Fourier Transform enables precision in diagnostics, enhances data analysis, and facilitates technological advancements in various research fields.

A key feature of this volume is its interdisciplinary approach, bridging fundamental mathematics with applied sciences and engineering. Chapters dedicated to signal processing discuss advanced audio and visual data analysis techniques, while those focused on spectroscopy and imaging illustrate how Fourier Transform aids in extracting critical information from complex physical systems. The book also delves into its applications in nanomaterials, where Fourier-based techniques play an essential role in characterizing structural and optical properties at the atomic level.

This work is intended for researchers, engineers, and graduate students seeking a deeper understanding of Fourier analysis and its applications beyond conventional signal processing. By bringing together leading experts in the field, *Beyond Signals – Exploring Revolutionary Fourier Transform Applications* provides a valuable resource for those interested in leveraging Fourier Transform methodologies for scientific innovation and technological progress.

We sincerely thank the contributors who have shared their expertise and insights and the editorial team for their meticulous efforts in assembling this volume. Special thanks to my beloved children, Arfa Bilal, Zainab Chaudhary, and Muhammad Musa, for their endless support and inspiration.

Dr. Muhammad Bilal Tahir
Associate Dean,
Faculty of Natural and Applied Sciences,
Institute of Physics,
Khwaja Fareed University of Engineering and Information Technology (KFUEIT),
Pakistan

Chapter 1

A ATR-FTIR Study of Extracted Bioapatite from Archaeological Bone

Antonio Martínez Cortizas and Olalla López-Costas

Abstract

Extracted bioapatite (BAp) from archaeological skeletal remains is used in studies reconstructing diet, mobility, and lifestyle of past human populations. Despite this, few investigations have researched extracted BAp, although some of them suggest that chemical extraction may alter BAp structure and composition. Here, we apply attenuated total reflectance-Fourier transform infrared (ATR-FTIR) spectroscopy on BAp extracted from archaeological bones: 29 human individuals of two medieval-postmedieval necropolises and 30 non-humans from Roman and medieval times, all from northwestern Spain. Statistical treatment includes principal component analysis (PCA) on (i) selected peaks (dPCA) and (ii) whole spectrum (tPCA). Extracted BAp shows the characteristic vibrations of the main components of a non-stoichiometric apatite, as well as other minor vibrations related to labile components. PCA results are also consistent in revealing the structural and compositional features of BAp, although tPCA results correlate better to BAp IR indices and properties than dPCA results. Significant differences in BAp spectral signal were found between human and non-human bone, collections (i.e., necropolises), human groups (non-adults vs. adults), human bone type (long bones vs. ribs), and non-human groups (bovine vs. ovicaprid). Thus, extracted BAp also retains crucial information that could be related to pre- and postmortem modifications of bone.

Keywords: ATR-FTIR, archaeological bone, bioapatite, yield, carbonate content, C stable isotopes

1. Introduction

Bioapatite (BAp) is the main mineral component of bone and other human skeletal remains and may represent up to 70% of the bone mass [1]. BAp is widely used in osteoarchaeological research to provide information on both ante- and postmortem characteristics. For example, the isotopic composition of BAp is used to reconstruct diet ($\delta^{13}C$ in carbonates), specifically to distinguish intake of marine/terrestrial resources and C_3/C_4 plants complementing the collagen isotopic study (e.g., [2]). The isotopic composition of BAp in teeth and/or bones ($\delta^{18}O$ in carbonates) also constitutes one of the most preferred approaches to paleomobility due to its relation to rainfall water (e.g., [3]), and it did not decrease in popularity despite the use of bone

BAp has recently been criticized due to potential postmortem alterations (e.g., [4]). Indeed, many studies are focused on understanding postmortem changes in bone (i.e., diagenetic alterations, e.g., [5–10]). Finally, in other studies, extracted BAp is used to address physiological changes related or not with disorders (e.g., [11, 12]) and to better understand changes in paleopathological bone composition, for example, if infectious diseases could interfere with ^{14}C dating [13].

Despite the large amount of literature that is based on extracted BAp, some investigations do not recommend the use of extracted BAp to research aspects related to mineral structure, degree of order, maturity, and crystallinity, as the chemical extraction may introduce significant modifications [1, 14]. They recommend the direct use of the spectral information obtained in bulk bone despite they also agree that overlapped vibrations from different components, as those arising from collagen, introduce uncertainty in the calculation of peak intensities, areas, and ratios (i.e., IR indices). In some cases, mathematical subtraction of the collagen spectrum—after multiplication by an *ad hoc* coefficient—is recommended to account for/correct the collagen signal [14]. This is also problematic, as it has already been shown that collagen undergoes diagenetic changes that result in decreased absorbance in the amide region (1900–1200 cm^{-1}) and increased absorbance in the proteoglycan region (1200–800 cm^{-1}) [15], both overlapping with the phosphate and carbonate absorbances. In fact, Grunewald et al. [1] already indicated that collagen subtraction must be made with the spectrum of the collagen extracted on the same bone sample. Some investigations use a combination of synthetic apatite samples and archaeological bone samples, but many of them rely on a small (3–10 samples) subset of archaeological bone samples, which may hamper the strength of the conclusions.

To our knowledge, not many systematic investigations have been performed on extracted BAp to determine the extent of the changes introduced by the extraction procedure. Here, we apply attenuated total reflectance-Fourier transform infrared (ATR-FTIR) to characterize the BAp from archaeological bone in order to address its composition and structure. We extracted BAp from 29 human individuals belonging to two medieval–postmedieval necropolises from northwestern (NW) Spain and 30 animal bones from Roman and medieval times, as well as two recent samples (one from cow and one from pig). Statistical treatment includes principal component analysis (PCA) on (i) selected peaks (based on the second derivative spectra, dPCA) and (ii) on whole spectrum (tPCA). Our aims are: (i) to determine the characteristics of the extracted BAp and compare them with previous investigations on non-extracted (i.e., bulk bone) BAp, (ii) relate the IR indices and principal components to chemical properties of BAp, and (iii) determine if the extracted BAp retains differences in composition between and within the collections analyzed: human vs. non-human, archaeological sites collections (both for humans and non-humans), different types of bone (ribs vs. long bone), human groups (i.e., non-adults vs. adults), and non-human group (i.e., bovids vs. ovicaprid).

2. Material and methods (skeletal collections, analytical procedures, and previous data)

2.1 Collections and samples

The studied collections belong to two archaeological areas placed in NW Spain. Human samples were obtained from the inside-wall area of Pontevedra, a medieval village by the sea. Skeletons were found in two funerary areas, one placed surrounding San Bartolomeu O Vello church (SB) and the other in Santa María A Maior church (STM), approximately 100 m from the other. Same type of soil and climatic conditions may have affected skeletons from SB and STM despite changes related to microscale taphonomy and weathering. Animal samples belonged to different archaeological surveys done in medieval Pontevedra. In addition, we have added another collection found in A Lanzada site (buried in sand dune), also by the sea and placed 20 km away. Human samples are presumably ascribed to late medieval–postmedieval period (most from thirteenth to fifteenth centuries). Animal samples from Pontevedra may have had the same chronology, while animal samples from Lanzada would be earlier (first to seventh centuries).

Samples were taken mostly from long bones using the remains of previous studies of collagen extraction for paleodiet published elsewhere [16–18]. Due to ethical reasons, we avoided new sampling since BAp extraction is a destructive method.

2.2 Bioapatite extraction and properties (yield, carbonate content, and $\delta^{13}C$)

BAp was extracted following Garvie-Lok et al. [19] protocol with modifications by López-Costas et al. [20]. Bone fragments were ultrasonic cleaned at least 5 times at 5 minutes each in ultrapure water type 1. Then, fragments were dried at 25 degrees all night and keep them in a low-humidity environment. Bone fragments were then finely milled using a mill Resch MM 301 and finishing the process with a hand mortar. Here, 250 mg of milled bone was placed in falcon tubes together with 10 ml of 2.5–3% aqueous sodium hypochlorite. Samples were slowly (10 rpm) and continuously homogenized using a rotor during four sets of 24 h each to completely remove the organic fraction. The samples were then rinsed five times with ultrapure water type 1 before adding 10 ml of 0.1 M of acetic acid and leaving them for 4 h at room temperature. The acid was rinsed again five times with ultrapure water, and samples were transferred to Eppendorf tubes in order to be frozen and freeze-dried. All processes took place at clean lab facilities of the research group EcoPast in the Universidade de Santiago de Compostela, by strict measurements of anti-contamination [18].

Carbon isotope ratios in carbonate from BAp were measured using a Europa Scientific 20–20 IRMS by adding phosphoric acid and measuring CO_2 by continuous flow-isotope ratio mass spectrometry (CF-IRMS) at Iso-Analytic Inc. Reference materials were used NBS-18, NBS-19 are distributed by the international atomic energy agency (IAEA) as inter-laboratory comparison standard materials as well as internal or external standards. Analytical error was ± 0.2‰ or less. Carbonate content (%) and $\delta^{13}C_{car}$ of human samples were previously published for paleodiet reconstruction [18]; the same data for animals has not been published.

Addressing the quality of BAp and especially the impact of diagenetic changes in isotopic fingerprints ($\delta^{13}C_{car}$) is key to prove the reliability of diet reconstruction. Bone BAp yield has been proposed as a reliable method for quality control in stable isotope analyses [21]. However, the range of good bone condition is quite wide (20–60%), and there is no clear consensus in literature about the relationship between BAp properties and yield. Carbonate content has been also used to detect the incorporation of exogenous carbonate (not removed by the acetic acid step) or the loss of structural carbonate [22]. However, beyond those facts, it is truly important to understand to what extent compositional/structural changes in BAp influence the isotopic signal, i.e., if there is a fractionation derived from postmortem modifications. If proven, the lack of fractionation will support the use of bone BAp to reconstruct aspects of the antemortem that are so important in osteoarchaeological research.

2.3 ATR-FTIR

The BAp samples were analyzed by attenuated total reflectance Fourier transform infrared spectroscopy (ATR-FTIR). We used Cary 630 spectrometer (Agilent Technologies) equipped with a diamond crystal. Spectra were acquired in the mid-infrared (MIR) region (4000–400 cm^{-1}) with a resolution of 4 cm^{-1}. Every sample spectrum was computed after background correction and 100 scans. Baseline correction, second derivative calculations, and calculation of peak areas were done using the spectroscopy add-on widgets of the Orange data mining software [23], while the R {*andurinha*] package was used to determine peak locations (i.e., wavenumber) [24].

We also calculated a number of IR indices which are assumed to reflect the abundance of the main BAp components and the degree of structural order and crystallinity:

- Carbonate relative to phosphate: C/Pi (using maximum peak intensities) and C/Pa (using peak areas) [1].
- Type of carbonate substitution: B-type (BPI, substitution for PO_4) and A-type (API, substitution for OH) [25].
- Relative abundance of the substitutions: C/C (A-type/B-type) [25].
- Presence of amorphous carbonate: AC (high crystalline apatite versus poorly crystalline apatite phosphates) [12].
- Degree of structural order and crystallinity of the BAp: IR splitting factor (IRSF), IRSF [25, 26], full width at half maximum (FWHM) of the main phosphate absorbance peak (FWHM) [1], and mineral maturity index (MMI) [27].

2.4 Statistical methods

Principal components analysis was performed on the correlation matrix and with varimax rotation on (i) selected peaks of the BAp spectra (dPCA) and (ii) the whole spectra in the fingerprint region (1700–430 cm^{-1}). In the second case, the spectra of the samples are in the columns and the wavenumbers on the rows, i.e., we use the transposed matrix (tPCA). This way of performing the PCA is efficient in decomposing the spectrum of every sample in a number of scores' spectra that are related to the compounds present in the samples. Loadings are assigned to the samples and reflect the proportion of the spectral variance of each sample that is related to each compound. This approach has proved to be efficient in extracting MIR spectral information on environmental samples [28, 29], and its application is described in detail elsewhere [30].

To check for significant differences between human and non-human bone BAp, archaeological sites, human groups (non-adults vs. adults), human bone type, and non-human group (bovids vs. ovicaprid), we use non-parametric statistics due to the relatively small number of samples in the groups. The Mann–Whitney U test was used when comparing two groups, and the Kruskal–Wallis test for more than two groups. Multilinear regression models were also computed in stepwise mode. All statistical analyses were performed with IBM SPSS Statistics v29.0.2.

3. Results

3.1 Bioapatite properties: Yield, carbonate content, and isotopic composition ($\delta^{13}C$)

Data on the BAp properties that are used to correlate with MIR data can be found in **Table 1**. Yield shows high average values for all samples, being above 40% except for the non-human bones of A Lanzada, which have low BAp recoveries. Yield is also low and variable for the fresh bone samples, possibly related with larger organic compounds content due to better preservation than archaeological samples (e.g., lipids). The average carbonate content in the archaeological samples (5.6–9.1%) is comparable to values found in previous investigations [1, 18, 31, 32]. Regarding C isotopic composition (**$\delta^{13}C$** in carbonate), results related to Pontevedra necropoleis were already published for paleodiet reconstruction [18], being variation in humans from STM larger than human individuals from SB. Those are compatible with diets rich in C_4 plants and marine resources, a fact that agrees with previous studies on collagen [16, 17].

3.2 ATR-FTIR of bioapatite extracted from archaeological skeletal remains

The spectra of the analyzed samples as well as the average second derivative spectrum can be found in **Figure 1**. The accompanying table in the figure provides the range, mode, bonds implicated and corresponding vibrational modes, based on literature [1, 14, 31–41]. The identified vibrations mostly correspond to the main components of the BAp, phosphate and carbonate, but vibrations from OH and adsorbed water are also identified.

For the phosphate group, the samples showed the vibrational modes of the apatitic environment, with the highest absorbance intensities corresponding to the ν_3 (1019 cm^{-1}) and the doublet of the ν_4 (600 and 559 cm^{-1}) modes. The ν_4 mode has also shoulders at 576 and 1030 cm^{-1}. Vibrational modes ν_1 (960 and 1113 cm^{-1}) and ν_2 (472 cm^{-1}) show moderate to low-intensity values. Low absorbances were found for peaks 615, 535 (non-apatitic HPO_4^{2-})518, and 498 cm^{-1} (ν_4PO_4).

As for the carbonate, absorbances of both B-type (1472, 1452, 1420, 1411, and 874 cm^{-1}) and A1-type (1549 cm^{-1}) substitutions (for PO_4 and OH, respectively) show moderate to low intensities, and correspond to the ν_2 and ν_3 vibrational modes

Site	Yield (%)	Carbonate (%)	$\delta^{13}C$ (%0)
H.SB	47.5 ± 10.1 (36.0–64.9)	6.8 ± 0.7 (5.6–7.6)	−11.70 ± 0.82 (−12.65 to −10.56)
H.STM	46.1 ± 13.6 (5.0–61.6)	7.1 ± 1.1 (4.8–9.1)	−12.60 ± 0.97 (−14.69 to −10.87)
F.LZ	5.8 ± 2.7 (3.0–12.6)	7.7 ± 0.87 (6.6–9.1)	−11.33 ± 1.04 (−12.43 to −10.47)
F.STM	50.5 ± 12.5 (17.8–64.4)	N.A.	N.A.
F.RB	4.7–20.0	N.A.	N.A.

H: human, F: non-human, SB: San Bartolomé, STM: Santa María, LZ: A Lanzada, RB: fresh bone, and N.A.: data not available.

Table 1.
Average, standard deviation (±), and range (in parenthesis) for the values of the properties of the extracted bioapatite samples.

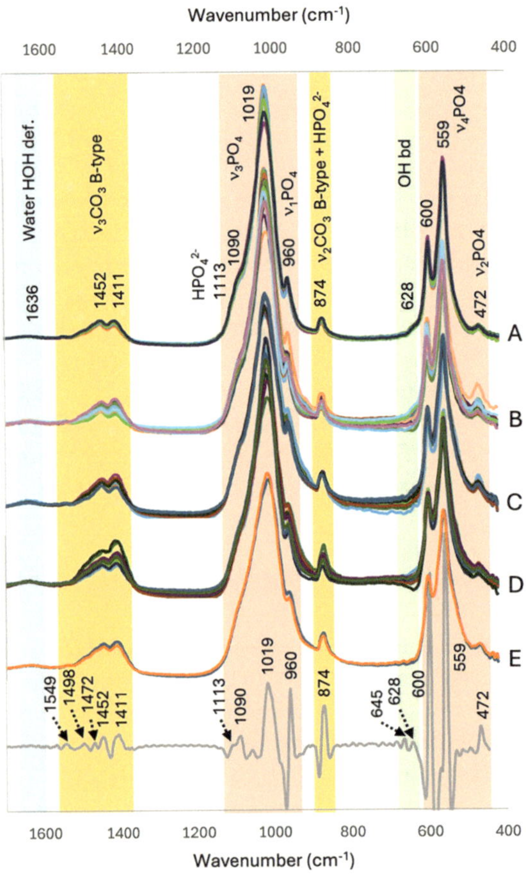

Range	Mode	Assignation
1111-1118	1113	HPO_4^{2-}
1090-1100	1090	ν_3PO_4
1030-1034	1030	ν_3PO_4
1012-1025	1019	ν_3PO_4
956-962	960	ν_1PO_4
615-621	615	Non-apatitic PO_4
600-602	600	ν_4PO_4
572-578	576	ν_4PO4
559-561	559	ν_4PO_4
526-537	535	Non-apatitic HPO_4^{2-}
510-510	518	ν_4PO_4
490-505	498	ν_4PO_4
470-473	472	ν_2PO_4
1539-1552	1549	CO_3 A1-type
1508-1519	1510	CO_3 A2-type
1493-1504	1498	CO_3 A2-type
1469-1472	1472	ν_2CO_3 B-type
1448-1456	1452	ν_3CO_3 B-type
1418-1422	1420	B-type + non-apatitic
1407-1415	1411	ν_3CO_3 B-type
870-876	874	ν_2CO_3 B-type + HPO_4^{2-}
639-654	645	OH bd
629-639	628	OH bd
1630-1650	1636	HOH deformation
684-695	688	H_2O libration mode
665-673	667	H_2O libration mode

Figure 1.
Mid-infrared spectra of the analyzed samples, grouped by archaeological collection, and (reversed) average second derivative spectrum for the fingerprint region (1900–450 cm^{-1}). A: human samples from San Bartolomé, B: human samples from Santa María, C: non-human samples from A Lanzada, D: non-human samples from Santa María, E: non-human fresh bone samples. The table to the right synthesizes the main vibrations identified, indicating the range and mode for the position of each peak and the corresponding bond and vibration mode.

of CO_3. The presence of labile, A2-type, carbonate is suggested by small peaks (1510 and 1498 cm^{-1}) in the second derivative spectrum (**Figure 1**).

A small shoulder in the 650–620 cm^{-1} region suggests the presence of hydroxyl ions typical of the BAp (i.e., hydroxylapatite); while low-intensity absorbances between 690 and 660 cm^{-1}, as well as 1636 cm^{-1}, correspond to libration and deformation modes of water.

3.3 IR indices of bioapatite

As indicated in the material and methods section, a number of IR indices were calculated based on the literature. These indices are assumed to reflect the abundance of carbonate relative to phosphate (C/Pi and C/Pa), the degree of B-type (BPI), and A-type (API) carbonate substitutions, their relative abundance (C/C), the presence of amorphous carbonate (AC), and the degree of structural order and crystallinity of the BAp (ISRF, FWHM, MMI). Values for these indices are represented in **Figure 2**.

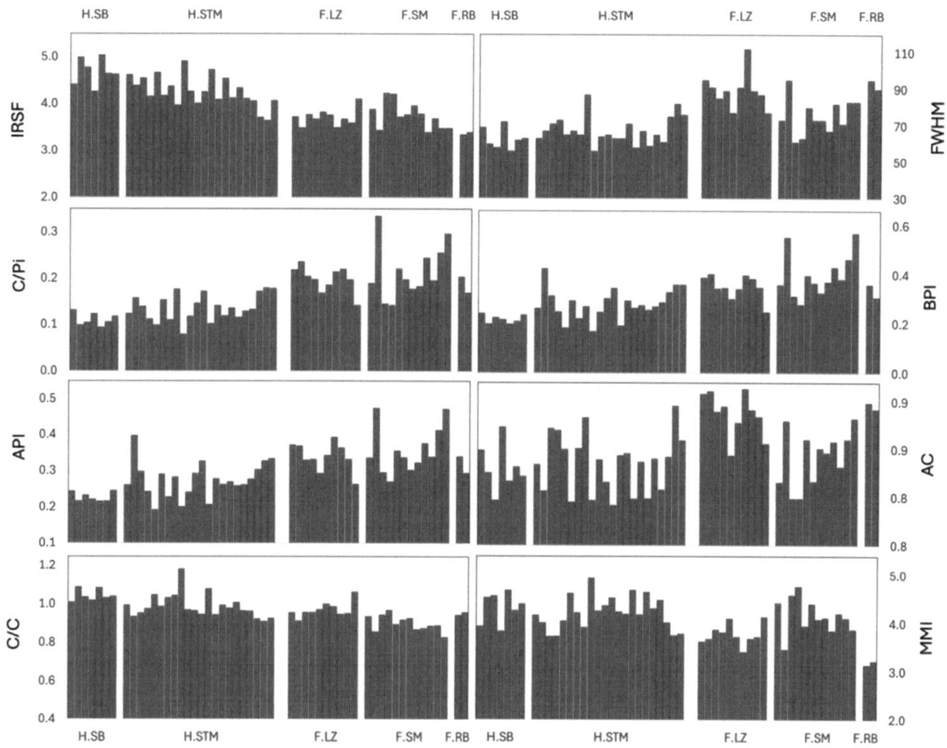

Figure 2.
Values of the IR indices calculated for the analyzed samples. C/Pa is not represented because it is highly correlated (r 0.99) to C/Pi—see below. H: human, F: non-human, SB: San Bartolomé, STM: Santa María, LZ: A Lanzada; RB: fresh bone.

Most of the indices show highly significant differences between human and non-human BAp compositional and structural properties. Human bones have higher values for IRSF ($p < 0.01$), MMI ($p < 0.01$), and C/C ($p < 0.01$), while values are lower for C/P (both C/Pi and C/Pa, $p < 0.01$), API ($p < 0.01$), BPI ($p < =0.01$), AC ($p < 0.01$), and FWHM ($p < 0.01$). The two necropolises also differ significantly: San Bartolomé has higher values for IRSF and the A/B ratio, but lower for C/P, API, and BPI than Santa María. The type of bone analyzed in the humans' sample only shows significant differences for C/C (long bones showing higher values than ribs, $p = 0.04$), and FWHM (long bones showing lower values than ribs, $p = 0.04$). While the human groups (non-adults, women, and men) only show significant differences for MMI ($p < 0.01$), with non-adults having lower values than adults (whether women or men).

As for the non-human bone samples, A Lanzada ones have higher values for C/C ($p < 0.01$), AC ($p < 0.01$), and FWHM ($p < 0.01$), but lower values than those of Santa Maria for MMI ($p < 0.01$). No differences were found when comparing bovid and ovicaprid bones. BAp extracted from fresh non-human bone shows values comparable to those of the BAp extracted from archaeological non-human bone (**Figure 2**).

Overall, the IR indices are highly correlated ($r > 0.7$, $r < -0.7$) between them (**Table 2**). The indices related to carbonate content (C/Pi, C/Pa) and type of substitution (BPI, API) show very large positive ($r > 0.9$) correlations between them and negative correlation with the A-type/B-type ratio (C/C; $r < -0.8$). The IR indices related to BAp structural order and crystallinity show high negative correlations (IRSF vs. FWHM and MMI with FWHM vs. MMI) or moderate positive correlations (IRSF vs. MMI). Also, these IR indices are highly correlated to the IR indices related to carbonate content and type of substitution: IRSF is negatively correlated to C/P (C/Pi and C/Pa), API, BPI, and AC, and positively correlated to C/C; while FWHM shows the opposite, but lower correlation coefficient values (**Table 2**).

3.4 Synthetic signals: Principal component analyses (dPCA and tPCA)

For the PCA on the standardized absorbance intensities (dPCA), we selected 26 peaks related to phosphate, carbonate, hydroxyl, and water vibrational domains. Five components account for 95.8% of the total variance. The first component, dCp1, accounts for 35.7% of the MIR spectral variance and shows large (>0.7) positive loadings for almost all the carbonate absorbances (1549, 1510, 1498, 1472, 1452, 1420, 1411, and 874 cm^{-1}), and moderate to high (0.5–0.7) loadings for water vibrations (1636 and 688 cm^{-1}) and non-apatitic PO_4 (615 cm^{-1}), and a moderate negative loading for ν_4PO_4 (1030 cm^{-1}) (**Table 3**). Thus, dCp1 reflects total carbonate content of BAp. The dCp1 scores show that almost all BAp samples from human bones have negative values, indicating lower carbonate content, while most of the non-human samples have positive scores indicating higher carbonate content (**Figure 3A**). The two samples from fresh non-human bone (F.RB) plot between the non-human archaeological samples and are also enriched in carbonate.

The second component, dCp2, accounts for 26.1% of the variance. Large positive loadings correspond to the main absorbances of the ν_1(960 cm^{-1}), ν_2 (472 cm^{-1}), and ν_4 (518 and 498 cm^{-1}) of phosphate, as well as non-apatitic HPO_4^{2-} (535 cm^{-1}) (**Table 4**). Water HOH deformation (1636 cm^{-1}), A-type carbonate substitutions (1549 cm^{-1}), and the ν_2CO_3 (B-type) have moderate positive loadings, while ν_3PO_4 mode (1019 and 1030 cm^{-1}) has moderate negative loadings. The component reflects changes between the non-apatitic (positive loadings) and apatitic phosphate environment. The dCp1/dCp2 projection provides a clear picture of BAp composition (**Figure 3A**). Most BAp samples from human bone tend to cluster together in the lower left quadrant of the dCp1/dCp2 projection, i.e., they have lower carbonate content but higher phosphate content (larger intensities of the main phosphate peak).

	MMI	C/Pi	C/Pa	API	BPI	C/C	AC	FWHM
ISRF	0.68	−0.86	−0.81	−0.79	−0.79	0.77	−0.63	−0.82
MMI		−0.58	−0.52	−0.52	−0.50	0.43	−0.91	−0.86
C/Pi			0.99	0.95	0.95	−0.79	0.56	0.77
C/Pa				0.94	0.96	−0.83	0.49	0.68
API					0.99	−0.81	0.45	0.70
BPI						−0.86	0.42	0.65
C/C							−0.31	−0.44
AC								0.87

Table 2.
Correlation (Pearson's correlation coefficient) between the calculated MIR indices of the human and non-human archeological bone samples.

WN		dCp1	dCp2	dCp3	dCp4	dCp5
1636	HOH deformation	0.67	0.65	−0.06	0.28	0.03
1549	CO_3 A1-type	0.69	0.64	−0.05	0.28	0.07
1510	CO_3 A2-type	0.90	0.37	−0.05	0.20	0.02
1498	CO_3 A2-type	0.93	0.32	0.02	0.16	−0.01
1472	ν_2CO_3 B-type	0.93	0.33	0.04	0.15	−0.04
1452	ν_3CO_3 B-type	0.91	0.37	0.09	0.13	−0.03
1420	B-type + non-apatitic	0.94	0.29	0.04	0.13	−0.06
1411	ν_3CO_3 B-type	0.94	0.30	0.05	0.11	−0.05
874	ν_2CO_3 B-type + HPO_4^{2-}	0.70	0.61	0.27	0.15	−0.07
688	H_2O libration mode	0.62	0.34	0.61	−0.22	−0.12
960	ν_1PO_4	0.47	0.79	0.16	0.17	−0.02
535	Non-apatitic HPO_4^{2-}	0.41	0.87	0.22	0.14	−0.04
518	ν_4PO_4	0.44	0.86	0.19	0.17	−0.07
498	ν_4PO_4	0.48	0.82	0.24	0.16	−0.07
472	ν_2PO_4	0.23	0.92	−0.07	0.24	−0.02
1030	ν_3PO_4	−0.52	−0.59	−0.40	0.21	−0.04
1019	ν_3PO_4	−0.48	−0.64	−0.39	−0.36	0.02
667	H_2O libration mode	0.47	0.25	0.74	−0.29	−0.11
645	OH bd	−0.14	0.15	0.94	−0.18	0.05
628	OH bd	−0.24	0.11	0.95	−0.12	−0.05
615	Non-apatitic PO_4	0.54	0.25	0.64	0.37	0.08
600	ν_4PO_4	0.28	0.11	0.80	0.41	0.14
559	ν_4PO_4	−0.16	−0.19	0.74	0.01	0.63
576	ν_4PO_4	0.42	0.08	0.69	0.48	−0.09
1113	ν_1PO_4	0.22	0.35	−0.29	0.85	0.06
1090	ν_3PO_4	0.22	0.32	0.05	0.91	−0.05

Table 3.
Loadings (dPCA) of the selected absorbances of the bioapatite in the five extracted principal components.

Only two samples from Santa María are slightly enriched in carbonates. For the non-human bones, all BAp samples from A Lazada (F.LZ, **Figure 3A**) have positive dCp2 scores (i.e., higher relative contribution of the non-apatitic PO_4), while samples from Santa María (F.STM, **Figure 3A**) have negative or slightly dCp2 positive scores (i.e., higher apatitic PO_4 content). Thus, dCp2 mainly accounts for differences in the non-human BAp and separates samples of the Roman necropolis (lower phosphate content) from those of the Medieval necropolis (higher phosphate content).

The third component, dCp3, accounts for 21.0% of the MIR spectral variance and shows large to moderate positive loadings for the OH bending vibrations (645 and 628 cm^{-1}), water libration mode (667 cm^{-1}), and ν_4PO_4 vibrations (600, 576, and 559 cm^{-1}). Positive loadings are also shown by the H_2O libration mode at 688 cm^{-1}. The component can be related to changes in the degree of hydroxylation (i.e., very

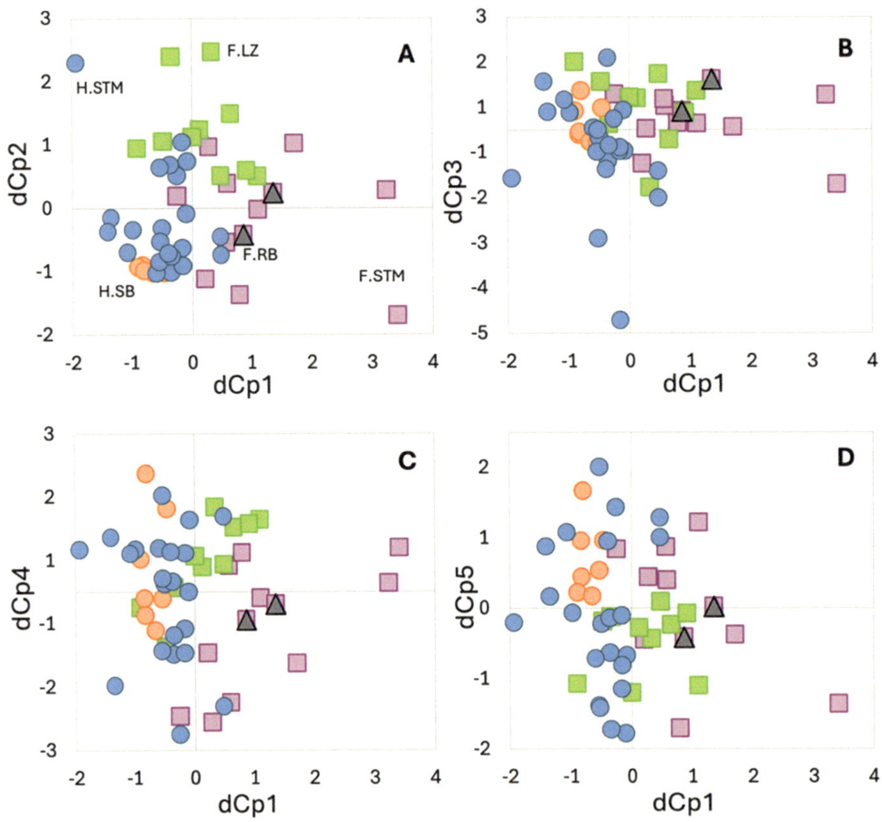

Figure 3.
Projections of the dPCA components extracted on the main peak absorbances of the extracted bioapatite.

	ISRF	MMI	C/Pi	C/Pa	API	BPI	C/C	AC	FWHM
dCp1	−0.73	−0.44	0.92	0.94	0.88	0.91	−0.79	0.37	0.56
dCp2	−0.44	−0.45	0.34	0.24	0.38	0.29	−0.09	0.47	0.65
dCp3	−0.23	−0.07	0.11	0.07	−0.18	−0.17	0.15	0.23	0.20
dCp4	−0.09	−0.61	0.07	0.08	0.01	0.01	−0.05	0.71	0.36
dCp5	0.31	−0.02	−0.02	0.01	−0.08	−0.09	0.17	−0.06	−0.06
tCp1	0.82	0.75	−0.80	−0.71	−0.74	−0.68	0.44	−0.76	−0.95
tCp2	−0.83	−0.75	0.79	0.70	0.73	0.67	−0.45	0.76	0.95
tCp3	−0.11	0.48	0.20	0.22	0.19	0.20	−0.19	−0.58	−0.21
tCp4	−0.34	−0.13	0.57	0.67	0.59	0.67	−0.71	0.09	0.09
tCp5	0.16	0.05	−0.05	−0.06	0.25	0.19	−0.03	−0.19	−0.03

Table 4.
Correlation (Pearson's correlation coefficient) of the dPCA and tPCA components with the IR indices used to characterize bioapatite degree of order, crystallinity, carbonate content, and maturity.

large loadings of the OH bending) of the BAp, amount of adsorbed water, and the ν_4PO_4 domain (related to BAp crystallinity). The scores of the component indicate that it is related to the specific nature of a few samples of the Santa María necropolis (**Figure 3B**) which have lower dCp3 scores.

The fourth component, dCp4, accounts for 11.2% of the MIR spectral variance. Two peaks in the high wavenumber domain [1] region, 1120–1080 cm^{-1} region, have large positive loadings in this component and correspond to non-apatitic, acidic phosphate (1113 cm^{-1}), and ν_4PO_4 (1090 cm^{-1}). The dCp4 scores show a relatively large spread both in human and non-human BAp samples suggesting variations in the abundance of non-stoichiometric apatite. Even more, the two non-human collections behave differently, with A Lanzada having higher scores than Santa María and the BAp of the fresh bone samples.

The fifth component, dCp5, accounts for 1.2% of the MIR spectral variance and only captures some secondary variations of the 559 cm^{-1} peak, which may be related to variations in the ν_4PO_4 vibrational domain. The scores also show a relatively large spread in both human and non-human samples, although San Bartolomé and A Lanzada are more homogeneous than the Santa María human and non-human samples.

It is interesting to note that in all dPCA projections, the non-human fresh bones samples plot between those of the non-human archaeological samples and close together, reflecting a quite homogeneous BAp.

BAp from human bones has significantly lower score values than BAp from non-human bones for the first three principal components (dCp1, dCp2, and dCp3), i.e., it has lower carbonate content, higher signal of the apatitic environment, and lower degree of hydroxylation. Between the human samples differences are only found for dCp2, with San Bartolomé bones having lower values (i.e., apatatic phosphate) than bones from Santa María. No differences exist between bone types and human groups for any of the components. As for the non-human bone, differences are significant for carbonate content (dCp1), apatitic environment (dCp2), and acidic phosphate content (dCp3), A Lanzada having higher values for the apatitic signal and acidic phosphate content, and lower for the carbonate content than bone from Santa María. No differences are found between bovine and ovicaprid bones.

To extract as much information as possible of the MIR spectral signal of the BAp samples, we also performed a principal components analysis on the transposed data matrix (tPCA), using the fingerprint region (1700–420 cm^{-1}) of the samples' spectra— contrary to bone MIR spectra, in extracted BAp the region between 4000 and 1900 cm^{-1} has very low absorbance and no distinctive peaks. Six principal components account for more than 99% of the MIR spectral variance of the samples. The first component, tCp1, shows a scores' spectrum record dominated by the phosphate maximum absorbance at 1019 cm^{-1} (ν_3PO_4 domain) and the doublet at 600 and 559 cm^{-1}(ν_4PO_4 domain), with a small contribution of the 1090 cm^{-1} (ν_3PO_4 domain) peak (**Figure 4**). The component seems to reflect the content of well-ordered/crystalline BAp. Samples from human bone have higher loadings than those from non-human bone, with variability within the two groups as Santa María necropolis has higher heterogeneity than San Bartolomé, and A Lanzada non-human samples have lower loadings than most of those from Santa María, but comparable values to the BAp of the fresh non-human bones.

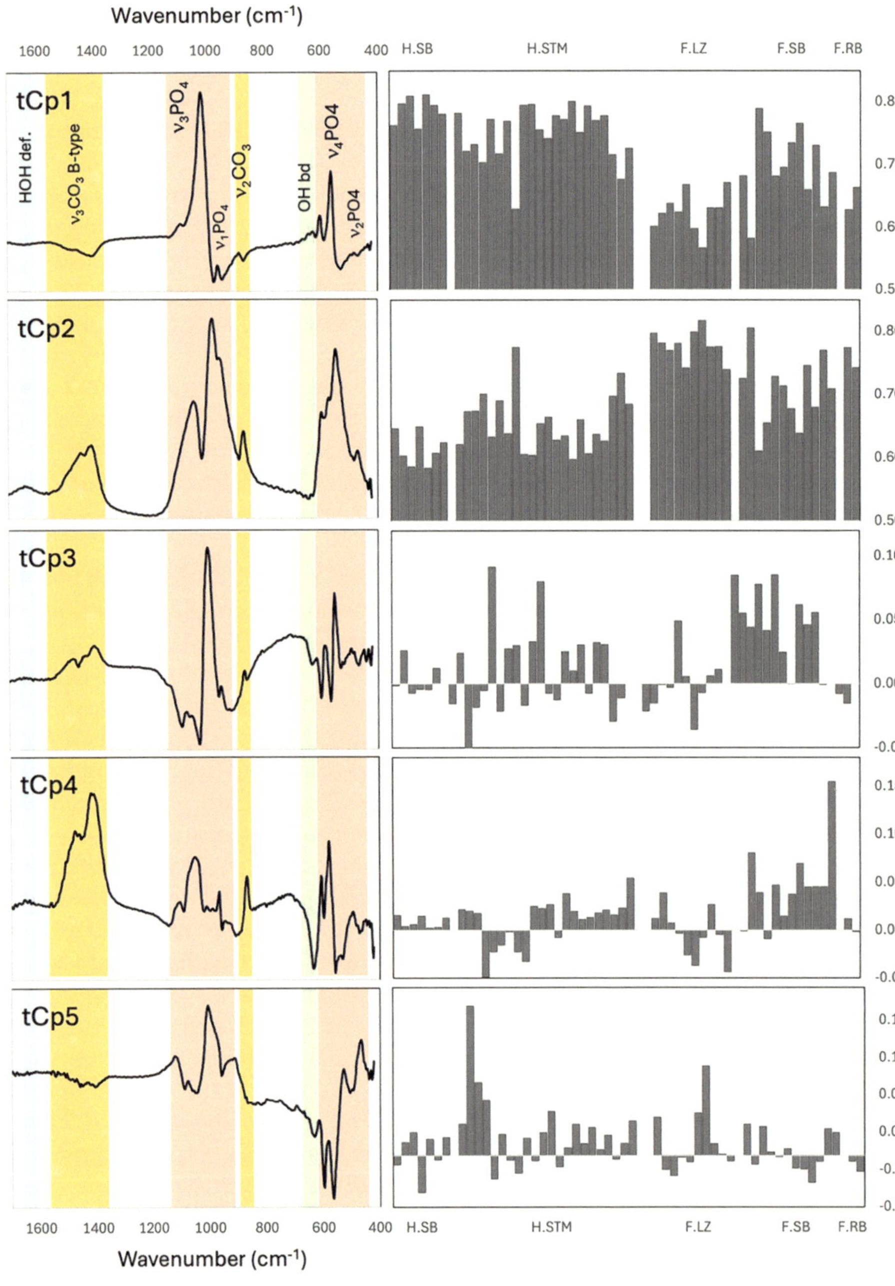

Figure 4.
Scores' spectra (left) and loadings of the bioapatite samples (right) resulting from the tPCA analysis.

The scores' spectrum of the second component, tCp2, shows large values for phosphate absorbances: maxima at 1044 and 952 cm^{-1} (both ν_1PO_4 mode) 574, 544, 511 (ν_4PO_4), and 470 cm^{-1} (ν_2PO_4); and for carbonate absorbances: 1500–1415 cm^{-1} (ν_3CO_3) and 874 cm^{-1}; with a small contribution of the HOH deformation mode of (adsorbed) water. The component reflects both the carbonate content of the BAp and changes in the width of the main phosphate peak, as well as secondary peaks in the short wavenumbers not related to tCp1. The distribution of the loadings in the samples is the opposite to that shown by tCp1: lower values of the BAp from human bone than those of non-human bone; much lower in San Bartolomé than in Santa María human bone, and higher in A Lanzada non-human bone compared to Santa María non-human bone. Again, A Lanzada has closer values to the fresh non-human bone.

The third component, tCp3, shows a scores' spectrum dominated by peaks of phosphates at 1001 cm^{-1} and 554 cm^{-1}, and a small contribution from the carbonate region (1500–1400 cm^{-1}) (**Figure 4**). This component also shows significant negative peaks of phosphate (1090, 1030, 602, and 567 cm^{-1}). It most likely reflects the relative contribution of the non-stoichiometric components (positive values) and stoichiometric components (negative values). The most remarkable feature in the loadings' distribution is the higher values shown by bones from Santa María necropolis, non-human ones in particular.

In the fourth component, tCp4, the MIR signal is dominated by carbonate, with peaks (1474 and 1403 cm^{-1}) at both sides of the typical carbonate doublet (1450 and 1415 cm^{-1}), and the peak at 870 cm^{-1}. Peaks on the phosphate regions are located at 1103, 1080, 1047, 967, 608, and 578 cm^{-1}. Negative peaks correspond to phosphate (1139, 958, 557, 5321, and 470 cm^{-1}) and OH (634 cm^{-1}). The component may thus be related to the abundance of carbonate B-type substitutions versus phosphate content and BAp hydroxylation. As with the previous component, positive loadings are shown by bones from Santa María necropolis with higher values in the non-human samples.

The fifth component, tCp5, reflects changes in the signal of the phosphate component. Positive scores occur at 1116, 1006, 973, 920, 527, and 466 cm^{-1}; while negative scores occur at 1090, 1044, 598, and 565 cm^{-1}. Moderate negative values also occur at 632 cm^{-1}, related to OH. The component is related to the balance between poorly crystalline apatite and crystalline, stoichiometric apatite. The tCp5 loadings show that this component is essentially related to a few (non-adult) human individuals of the Santa María Necropolis.

The loadings of the tPCA components exhibit significant differences between human and non-human bone BAp for tCp1, tCp2, and tCp3. Humans have a higher content of well-ordered/crystalline apatite (tCp1), lower carbonate and less ordered apatite, and lower non-stoichiometric components than non-human bone BAp. The only significant difference between human necropolises was found for tCp2, with higher values (i.e., less ordered apatite) in Santa María than in San Bartolomé. No significant differences are observed for type of bone or human group. In the non-human bones, four tPCA components have significant differences: BAp from A Lanzada bones has lower content of well-ordered/crystalline apatite and B-type carbonate substitutions, but higher carbonate content and less ordered apatite.

The two PCA approaches followed here seem to be consistent in showing the characteristics of the extracted BAp. In fact, almost all components are highly correlated: dCp1 is positively correlated to tCp4 ($r = 0.78$) and tCp2 ($r = 0.54$) and negatively correlated to tCp1 ($r = 0.55$); dCp2 is positively correlated to tCp2 ($r = 0.76$) and negatively correlated to tCp1 ($r = -0.77$); dCp3 is negatively correlated to tCp5 ($r = -0.89$); and dCp4 is negatively correlated to tCp3 (−0.85). dCp5 is the only dPCA component that shows no significant correlation to the tPCA components.

3.5 Relation between PCA signals, MIR indices and bioapatite properties

Apart from the consistency in the PCA approaches, another line of validation is to compare (i.e., correlate) them to the IR indices widely used in the literature, and to BAp properties relevant to osteoarchaeological research. Highly significant correlations are observed between the PCA components and the IR indices, except for dCp3 (i.e., hydroxylation), dCp5 (ν_1PO_4 domain), and tCp5 (i.e., non-stoichiometric apatite) (**Table 4**). dCp1 (carbonate) is highly correlated to the carbonate ratios (positive: C/P, API, BPI; negative: C/C) and the IRSF, and moderately correlated to FWHM. dCp2 (non-apatitic vs. apatitic phosphate) is moderately correlated to FWHM and dCp4 (non-stoichiometric) is highly correlated to AC and moderately (negative) correlated to MMI.

Of the tPCA, the first two are highly, and inversely, correlated so the correlation pattern with the IR indices is the opposite. tCp1 shows positive and high correlation with IRSF, and MMI, and negative correlation to C/P, API, BPI (moderate), and AC. tCp3 is negatively correlated to AC; and tCp4 is highly (negative) correlated to C/C, and moderately (positive) correlated to C/P, API, and BPI. Overall, the structure of correlation is stronger for the tPCA than for the dCPA, although the latter correlates better to the carbonate indices and the former with the apatite structural order and crystallinity.

Table 5 synthesizes the correlations between the IR indices and the principal components with bioapatite yield, carbonate content, and carbon isotopic composition (**δ^{13}**C). Bioapatite yield is positively (moderate) correlated to tCp1 and negatively correlated (moderate) with FWHM (negative), and dCp2 (negative). Carbonate content shows much higher correlation, the single largest one with the C/C ratio (negative),

	Yield (%)	**Carbonate (%)**	**δ^{13}C (%0)**
ISRF	0.46	−0.68	−0.20
MMI	0.49	−0.39	−0.16
C/Pi	−0.37	0.66	0.37
C/Pa	−0.28	0.69	0.36
API	−0.34	0.61	0.33
BPI	−0.26	0.69	0.27
C/C	0.06	−0.84	0.00
AC	−0.48	0.29	0.20
FWHM	−0.65	0.39	0.34
dCp1	−0.21	0.68	0.38
dCp2	−0.52	0.09	0.19
dCp3	−0.35	0.04	0.17
dCp4	−0.09	0.13	0.01
dCp5	0.04	−0.34	0.17
tCp1	0.67	−0.40	−0.36
tCp2	−0.67	0.43	0.35
tCp3	0.06	0.13	0.09
tCp4	0.24	0.54	0.00
tCp5	0.04	−0.15	0.03

Table 5.
Correlation (Pearson's correlation coefficient) of the dPCA and tPCA components, and the IR indices with the chemical properties of the bioapatite.

	Models	FWHM	API	BPI	C/C	AC	dCp1	dCp2	dCp3	tCp1	tCp2	tCp3	tCp4	tCp5	error	R
Yield	All		−4.07	4.99	0.49			0.66		1.35					11.7	0.85
	IR indices	−1.21			−0.37	0.44									14.3	0.75
	dPCA							−0.49	−0.31						16.9	0.61
	tPCA									0.68			0.27		14.6	0.72
Carbonate	All				−0.88							0.19		−0.23	0.47	0.90
	IR indices				−0.84										0.56	0.84
	dPCA						0.68								0.76	0.68
	tPCA									3.21	3.76	0.27	0.60		0.58	0.84

Table 6.
Multivariate linear regression models for bioapatite yield and carbonate content: normalized regression coefficients, error of the estimate, and regression coefficient (R). Regression was performed in stepwise mode using all IR indices and PCA components together (All), and separately for the IR indices, the dPCA, or the tPCA components.

but positive moderate correlations are also observed for C/P, API, BPI, and dCp1. No clear correlation is found between carbon isotopic composition and any of the indices or principal components.

The moderate to low correlation of the IR indices and PCA components with BAp properties suggest, as expected, that they are not dependent on only one aspect of the BAp structural characteristics for all the human and non-human samples analyzed here. Thus, we performed a multivariate, linear, regression analysis (MVLR) to test if a combination of BAp indicators is efficient in fitting the observed values of BAp yield, carbonate content and C isotopic composition. In this exercise, we used stepwise regression to determine the minimum number of variables needed to model the properties. We calculated regression models for all variables together (IR indices and PCA components), and separately (IR indices, dPCA components and tPCA components).

A synthesis of the models can be found in **Table 6**. In line with the low correlation found between the C isotopic composition and the IR indices and PCA components, no significant MVLR models were found for this property. For BAp yield, the best model (R 0.85) included carbonate IR indices (API, BPI, C/C) and PCA components (dCp1 and tCp1). Based on the normalized regression coefficients (**Table 6**), higher yields are correlated to higher B-type carbonate content, phosphate content, and better ordered/higher crystallinity of BAp. The IR indices and tPCA components alone also produce high correlation models, but lower than the combination of variables. While the dPCA components showed a modest fitting ability for yield.

As for carbonate content, a combination of the C/C index, tCp3, and tCp5 resulted in a regression model with a very high correlation (R 0.90) and low error of the estimate (5–9% of the maximum and minimum observed carbonate content) (**Table 6**). Higher carbonate contents are associated to lower A-type carbonate substitutions, higher content of non-stoichiometric components (tCp5), and lower content of poorly crystalline components. High correlation models were also found when using the IR indices and the tPCA components. As it was found for BAp yield, the dPCA components provided a moderate fitting model for carbonate content.

4. Discussion

4.1 Structural properties of the extracted bioapatite

One of the aims of this investigation was to determine the structural characteristics of the extracted BAp, as it regards to the MIR signal. Three main lines of evidence can be used to determine the extent of similarity in the characteristics of extracted BAp in comparison with those inferred from bulk bone: (i) the vibrational characteristics shown the MIR spectrum; (ii) the values of the IR indices typically used to infer aspects related to mineral maturity, order of the apatite lattice, and crystallinity; and (iii) relationship to relevant BAp properties, as BAp yield, carbonate content, and carbon isotopic composition ($\delta^{13}C$).

The spectra of the BAp extracted from human and non-human bones show the characteristic phosphate vibrational modes: ν_1 (960 cm^{-1}), ν_2 (472 cm^{-1}), ν_3 (1019 cm^{-1}), and ν_4 (the 600 and 559 doublet, and shoulders at 1030 and 576 cm^{-1}); the carbonate vibrational modes: ν_2 (870 cm^{-1}), and ν_3 (doublet at 1411 and 1452 cm^{-1}); a vibration related to OH (small shoulder at 628 cm^{-1}), and the deformation mode of adsorbed water (small shoulder at 1636 cm^{1}). Much less obvious, but detected by the second derivative, smaller absorbances were found also for phosphate: 1113, 615, and 535 cm^{-1}, related to HPO_3^{2-} ions, and 518 and 498 cm^{-1} that correspond to the PO_4 ν_4 mode; carbonate: 1472, 1452, and 1420 cm^{-1}, of the B-type substitutions, 1549 cm^{-1} of the A-type substitutions, and at 1510 and 1498 cm^{-1} that can be associated to A2-type carbonate (i.e., labile CO_3 ions). No evidence was found for secondary calcite (i.e., no peak at 712 cm^{-1}). Thus, in this regard, the extracted BAp shows all the characteristic components of non-stoichiometric hydroxylapatites typical of bone (e.g., [37] and references therein).

The range of values for IRSF (3.40–5.04) and MMI (3.41–4.94) in archaeological bone BAp are higher than those obtained for BAp of non-human fresh bone (3.36–3.40 and 3.14–3.22 respectively). For the other indices, the fresh bone values (C/Pi 0.17–0.20; C/Pa 0.16–0.18; API 0.29–0.34; BPI 0.31–0.36; A/B 0.94–0.94; AC 0.91–0.92; FWHM 90–95) are well within the range of archaeological samples (C/Pi (0.08–0.33), C/Pa (0.09–0.31), API (0.19–0.47), BPI (0.17–0.51), C/C (0.83–1.18), AC (0.81–0.93), and FWHM (56–112) for IRSF and MMI. Thus, mineral maturity and crystallinity of BAp from archaeological bone seem to be higher than those of fresh bone, which is in agreement with previous studies (e.g., [14]). The values for the other indices fall within the range found in previous investigations of archaeological bone [10, 14, 25, 42–48]. They are also similar to values found for medieval collections of NW Spain [18], although in other studies in the same area values found for C/P and BPI are higher and values of MMI lower than those of the present study [49]. The consistency between the IR indices, i.e., the large and significant correlations between most of them (**Table 2**) also supports the consistency of the spectral information of the extracted BAp. As crystallinity/lattice order (IRSF, FWHM) and mineral maturity (MMI) increase, the proportion of B-type carbonate substitutions also increases, but the carbonate-to-phosphate ratios decrease.

Regarding the third line of evidence, the correlation between the spectral properties of the extracted BAp is in general moderate, except for the C/C ratio that shows that total carbonate content increases as the relative content in B-type CO_3 substitutions increase in relation to A-type substitutions. The MVLR models show a good fitting for BAp carbonate content (**Table 6**) and suggest that it is the result of complex interactions. For example, up to four tPCA components are necessary to obtain a good fitting, and these components are related to the phosphate content and crystallinity (tCp1), carbonate content and apatite lattice disorder (tCp2), relative contribution of non-stoichiometric/stoichiometric components, and predominance of B-type CO_3 substitutions. Interestingly enough, no correlation or MVLR model accounted for a significant proportion of the C isotopic composition of the samples, indicating that changes in carbonate content, whether substituting for PO_4 or OH, or as labile CO_3, do not affect the isotopic signature, which may correspond to that of the individual when alive. In a previous investigation, we also showed that collagen C isotopic composition was independent of collagen diagenetic transformations [15]. Thus, at least for the collections studied by us, compositional and structural changes undergone by the organic and the inorganic components of the bone do not seem to affect the C isotopic composition.

4.2 Archeological implications

Differences were observed in BAp structure between long and short bones, specifically between femur and ribs. Long bones BAp shows a higher degree of structural order and maturity, as well as more CO_3 for OH substitution. Traditionally,

Osteoarcheology has assumed that femora display slower turnover rates than other bones, being a large part formed during adolescence [50]. Therefore, antemortem differences related to bone turnover could explain the observed differences in BAp maturity. Despite the promise of this assumption, non-adults also display similar differences, and their youngness makes the turnover hypothesis less likely. In addition, a recent study suggested that rib bone remodeling could be also a slow process [51]. Both bone types, ribs and femora, are traditionally analyzed in isotopic studies. Femora are normally selected in those studies performed in collections with poor preservation or in comingle remains. In a previous research regarding elemental analyses, we found that ribs were more altered by diagenesis than femur or crania [52], as also observed by other authors [53, 54]. In this case, changes in structural order could be here related to diagenetic modifications after inhumation. However, antemortem variation cannot be ruled out.

The researched collection seems also to play an important role in the BAp composition. LZ collection, composed in this case by fauna, shows higher OH substitution, lower amorphous carbonate and higher structural order, but lower maturity than STM. LZ is a site by the sea and fauna were found in Arenosol (i.e., sand dune) with biogenic carbonate, a soil environment that promotes less intense diagenetic changes [52]. The LZ maturity is more similar to the fresh bone; therefore, the soil environment, less aggressive to skeletal remains, could be behind the observed differences. In the same way, the BAp from humans collected in SB site displays higher crystallinity and has lower carbonate (in relation to phosphate) than humans from STM site. Both parameters covary in our dataset. SB humans are also more homogeneous, probably due to the fact that all were found in the same survey (S.16), while STM humans were excavated surrounding the church in areas with different soil conditions. The relationship between the BAp parameters and indices and the site the bones belonged to is a promising tool for museums and old collections, when the information regarding location is lost. A wider study using skeletons from different soil conditions could help to address the potential of ATR-FTIR in reconstructing diagenetic conditions of the collection when this information is lost.

Finally, the bioarchaeological characteristic shows that the larger differences in BAp spectral (ATR-FTIR) properties is the species, non-human vs. human. Faunal and human BAp were different in all IR indices. Human samples present higher values in several indices related to the structure of BAp including crystallinity (IRSF), maturity (MMI), and relative proportion of carbonate substitution type (C/C). This fact is persistent independently of the age of the individual (human non-adults also present differences with fauna) or the collection they belong to (LZ and STM non-human samples were different from human ones). In addition, human samples seem to have lower carbonate content (lower C/P, API, BPI), a relatively higher amount of B-type carbonate (lower C/C), higher content of amorphous carbonate (lower AC), and more crystalline structure (higher IRSF and MMI). All these facts point to a structural and compositional difference between non-human and human BAp. This is an interesting aspect to be applied in archaeological and forensic sciences, since in both cases being able to differentiate human bones is key to several studies. In a chapter related to the field study in Forensic Anthropology, we recently indicated the importance of rapid identification of human from non-human bones in order to stop or continue the search for missing persons [55]. ATR-FTIR can be used as a quick and relatively cheap option to address the species of bone fragment. An additional note is related to the already mentioned lack of correlation between BAp structure and $\delta^{13}C$, indicating that observed changes do not seem to influence the isotopic fingerprint that is normally used to reconstruct diet.

5. Concluding remarks

When compared with published data on BAp structural and chemical properties inferred from bulk bone ATR-FTIR analyses, our data on extracted BAp shows a large similarity. The vibrational modes of the major and minor components of BAp are present in the spectra of the extracted BAp and the values obtained for the IR indices fall within the range of those published for non-extracted BAp. It is also remarkable that correlation to properties like yield and carbonate content show similar trends to bulk bone BAp.

Regarding the application of chemometric statistical techniques (i.e., PCA), our results indicate that both dPCA applied on a selection of bands and the tPCA applied on the whole fingerprint region produce consistent, highly correlated synthetic variables. These variables are highly correlated with the IR indices and can be used to model BAp carbonate content and BAp yield. But the tPCA provides better correlations and has the advantage that the components are interpreted based on scores' spectra, which are more intuitive and provide better spectral information.

Extracted BAp is also shown here as a valuable tool to explore differences between bone types, conditions on site and, more important, to distinguish human from non-human samples. Open debates are still ongoing on how to specifically undertake antemortem vs. postmortem influence in archaeological bone composition, BAp study is opening an interesting world of opportunities in both Archeological and Forensic Sciences.

Due to the complexities of the BAp composition, we suggest that future studies should be based on a combination of techniques (FTIR, XRD, elemental, and isotopic analyses) and must include a larger number of samples from different soil/geochemical environments, if we are to be able to separate antemortem from postmortem transformations of the BAp structure and composition.

Finally, we recommend the use of extracted BAp from bone in osteoarchaeological studies, since the extraction produces no-change in its structure and composition. However, the consistency with the type of analyzed bone is important since we have detected slight intra-skeletal variations. The lack of isotopic fractionation ($\delta^{13}C$ in carbonates) in relationship to postmortem alteration of BAp composition and structure supports the use of isotopic composition to reconstruct diet, even in cases of poorly preserved remains. We also think that this work opens an unexpectedly very interesting field in zooarchaeological studies since our results point out that BAp fingerprints the species. Even, forensic applications could be implemented to quickly discard small bone fragments as non-human.

Acknowledgements

This research was funded by the European Union (ERC Consolidator Grant, PollutedPast, 101087832). Views and opinions expressed are however those of the author(s) only and do not necessarily reflect those of the European Union or the European Research Council. Neither the European Union nor the granting authority can be held responsible for them. This project was also funded by Grupos de Referencia Competitiva (ED431C 2021/32) by Xunta de Galicia. OLC is funded by Ramón y Cajal 2020 (RYC2020-030531-I). The authors would like to thank Ainé Francos Golán, Marta Colmenares Prado, Clara Veiga Rilo, and Noelia Rivero Chaver for their help during bioapatite extraction. Extended thanks to Patrimonio Histórico da Xunta de Galicia and Museo Provincial de Pontevedra for granting access to samples and skeletons.

Author details

Antonio Martínez Cortizas[1,2]* and Olalla López-Costas[3,4]

1 EcoPast (GI-1553), CRETUS, Department of Edafoloxía e Química Agrícola, Facultade de Bioloxía, Universidade de Santiago de Compostela, Spain

2 Bolin Centre for Climate Research, Stockholm University, Sweden

3 EcoPast (GI-1553), CRETUS, Area of Archaeology Department of History, Universidade de Santiago de Compostela, Spain

4 Archaeological Research Laboratory, Stockholm University, Sweden

*Address all correspondence to: antonio.martinez.cortizas@usc.es

References

[1] Grunenwald A, Keyser C, Sautereau AM, Crubézy E, Ludes B, Drouet C. Revisiting carbonate quantification in apatite (bio) minerals: A validated FTIR methodology. Journal of Archaeological Science. 2014;**49**:134-141. DOI: 10.1016/j.jas.2014.05.004

[2] Ambrose SH, Norr L. Experimental evidence for the relationship of the carbon isotope ratios of whole diet and dietary protein to those of bone collagen and carbonate. In: Lambert JB, Grupe G, editors. Prehistoric Human Bone—Archaeology at the Molecular Level. New York: Springer-Verlag; 1993, 1993. pp. 1-37

[3] Prowse TJ, Schwarcz HP, Garnsey P, Knyf M, Macchiarelli R, Bondioli L. Isotopic evidence for age-related immigration to Imperial Rome. American Journal Physical Anthropology. 2007;**132**:510-519

[4] Lightfoot E, O'Connell TC. On the use of biomineral oxygen isotope data to identify human migrants in the archaeological record: Intra-sample variation, statistical methods and geographical considerations. PLoS One. 2016;**11**:e0153850

[5] Keenan SW. From bone to fossil: A review of the diagenesis of bioapatite. American Mineralogist. 2016;**101**:1943-1951. DOI: 10.2138/am-2016-5737

[6] Keenan S, Engel AS. Early diagenesis and recrystalization of bone. Geochimica et Cosmochimica Acta. 2017;**196**:209-223. DOI: 10.1016/gca.2016.09.033

[7] Kendall C, Eriksen AMH, Kontopoulos I, Collins MJ, Turner-Walker. Diagenesis of archaeological bone and tooth. Palaeography Palaeoclimatology Palaeoecology. 2018;**491**:21-37. DOI: 10.16/j.palaeo.2017.11.041

[8] Pucéat E, Reynard B, Lécuyer C. Can crystallinity be used to determine the degree of chemical alteration of biogenic apatites? Chemical Geology. 2004;**205**:83-97

[9] Scaggion C, Dal Sasso G, Nodari L, Pagani L, Carrara N, Zotti A, et al. An FTIR-based model for the diagenetic alteration of archaeological bones. Journal of Archaeological Science. 2024;**161**:105900. DOI: 10.1016/j.jas.2023.105900

[10] Wright LE, Schwarcz HP. Infrared and isotopic evidence for diagenesis of bone apatite at dos Pilas, Gutemala: Paleodietary implications. Journal of Archaeological Science. 1996;**23**:933-944

[11] Isaksson H, Turunen MJ, Rieppo L, Saarakkala S, Tamminen IS, Rieppo J, et al. Infrared spectroscopy indicates altered bone turnover and remodeling activity in renal osteodystrophy. Journal of Bone and Mineral Research. 2010;**25**:1360-1366. DOI: 10.1002/jbmr.10

[12] Zuluaga-Morales JS, Bolaños-Carmona MV, Cifuentes-Jiménes C, Álvarez-Lloret P. Chemical, microstructural and morphological characterisation of dentine caries simulation by pH-cycling. Minerals. 2022;**12**:5. DOI: 10.3390/min12010005

[13] Nagy G, Lorand T, Patonai Z, Motsko G, Bajnoczky I, Marcsik A, et al. Analysis of pathological and non-pathological human skeletal remains by FT-IR spectroscopy. Forensic Science International. 2007;**175**:55-60. DOI: 10.16/j.forsciint.2007.05.008

[14] Dal Sasso G, Asscher Y, Angelini I, Nodari L, Atioli G. A universal curve of apatite crystallinity for assessment of bone integrity and preservation. Scientific Reports. 2018;**8**:12025. DOI: 10.1038/s41598-018-3064-z

[15] Martínez CA, López-Costas O. Linking structural and compositional changes in archaeological human bone collagen: And FTIR-ATR approach. Scientific Reports. 2020;**10**:17888. DOI: 10.1038/s41598-020-74993-y

[16] López-Costas O, Müldner G. Fringes of the empire: Diet and cultural change at the Roman to post-Roman transition in NW Iberia. American Journal of Physical Anthropology. 2016;**161**:141-154

[17] López-Costas O, Müldner G. Boom and bust at a medieval fishing port: Dietary preferences of fishers and artisan families from Pontevedra (Galicia, NW Spain) during the late medieval and early modern period. Archaeological and Anthropolical Science. 2019;**11**:3717-3731

[18] Veiga-Rilo C, Martínez CA, López-Costas O. Biting into the truth: Connecting oral pathology and stable isotopes through the paradigmatic example of a hyper-specialized marine diet in medieval Pontevedra (NW Iberia). Archaeological and Anthropological Sciences. 2024;**16**:49. DOI: 10.1007/s12520-024-01956-z

[19] Garvie-Lok SJ, Varney TL, Katzenberg MA. Preparation of bone carbonate for stable isotope analysis: The effects of treatment time and acid concentration. Journal of Archaeological Science. 2004;**31**:763-776

[20] López-Costas O, Müldner G, Lidén K. Biological histories of an elite: Skeletons from the Royal Chapel of Lugo cathedral (NW Spain). International Journal of Osteoarchaeology. 2021;**31**:941-956

[21] Chesson LA, Beasley MM, Bartelink EJ, Jans MM, Berg GE. Using bone bioapatite yield for quality control in stable isotope analysis applications. Journal of Archaeological Science: Reports. 2021;**35**:102749

[22] Vaiglova P, Lazar NA, Stroud EA, Loftus E, Makarewicz CA. Best practices for selecting samples, analyzing data, and publishing results in isotope archaeology. Quaternary International. 2023;**650**:86-100

[23] Demsar J, Curk T, Erjavec A, Gorup C, Hocevar T, Milutinovic M, et al. Orange: Data mining toolbox in python. Journal of Machine Learning Research. 2013;**14**:2349-2353

[24] Álvarez Fernández N, Martínez Cortizas A. Andurinha: make spectroscopic data processing easier. R package version 0.0.2. 2020. Available from: https://CRAN.R-project.org/package_andurinha

[25] France CA, Sugiyama N, Aguayo E. Establishing a preservation index for bone, dentin, and enamel bioapatite mineral using ATR-FTIR. Journal of Archaeological Science: Reports. 2020;**33**:102551. DOI: 10.1016/j.jasrep.2020.102551

[26] Weiner S, Bar-Yosef O. States of preservation of bones from prehistoric sites in the near east: A survey. Journal of Archaeological Science. 1990;**17**:187-196. DOI: 10.1016/0305-44305(90)090058-D

[27] Farlay D, Panczer G, Rey C, Delmas PD, Boivin G. Mineral maturity and crystallinity index are distinct characteristics of bone mineral. Journal of Bone and Mineral Metabolism. 2010;**28**:433-445. DOI: 10.1007/s009-01467-7

[28] Kylander ME, Martínez CA, Sjöstrom JK, Galing J, Gyllencreutz R,

Bindler R, et al. Storm chasing: Tracking Holocene stormines in southern swede using mineral proxies from inland and coastal peat bogs. Quaternary Science Reviews. 2023;**299**:107854. DOI: 10.1016/j.quascirev.2022.107854

[29] Martínez CA, López-Merino L, Silva-Sánchez N, Sjöstrom JK, Kylander ME. Investigating the mineral composition of peat by combining FTIR-ATR and multivariate analysis. Minerals. 2021;**11**:1084. DOI: 10.3390/min11101084

[30] Martínez Cortizas A, Traoré M, López-Costas O, Sjöström JK, Kylander M. Beyond the obvious: Exploring peat vibrational spectroscopy (FTIR-ATR) data using principal comonents analysis on transposed data matrix (tPCA), Store Mosse bobg (Sweden). In: Núñez Delgado A, editor. Frontier Studies in Soil Science. Cham, Switzerland: Springer; pp 217-247

[31] Dal Sasso G, Lebon M, Angelini I, Maritan L, Usai D, Artioli G. Bone diagenesis variability among multiple burial phases at Al Khiday (Sudan) investigated by ATR-FTIR spectroscopy. Palaeography Palaeoclimatology Palaeoecology. 2016;**463**:168-179. DOI: 10.1016/j.palaeo.2016.10.005

[32] Kontopoulos I, Presslee S, Penkman K, Collins MJ. Preparation of bone powder for FTIR-ATR analysis: The particle size effect. Vibrational Spectroscopy. 2018;**99**:167-177. DOI: 10.1016/j.vibspec.2018.09.004

[33] Asscher Y, Weine S, Boaretto E. Variations in atomic disorder in biogenic carbonate hydroxyapatite using the infrared spectrum grinding curve method. Advanced Functional Materials. 2011;**21**:3308-3313. DOI: 10.1002/adfm.201100266

[34] Balan E, Delattre S, Roche D, Segalen L, Morin G, Guillaumet M, et al. Line-broadening effects in the powder infrared spectrum of apatite. Physics and Chemistry of Minerals. 2010;**38**:111-112. DOI: 10.1007/s00269-010-0388-x

[35] Berzina-Cimdina L, Borodajenko N. Research of calcium phosphates using Fourier transform infrared spectroscopy. In: Infrared Spectroscopy. London, UK: IntechOpen; 2012. DOI: 10.5772/36942

[36] Brangule A, Gross KA. Importance of FTIR spectra deconvolution for the analysis of amorphous calcium phosphates. IOP Conference Series: Materials Science and Engineering. 2015;**77**:012027

[37] Figueiredo MM, Gamelas JAF, Martins AG. Characterization of bone and bone-based graft materials using FTIR spectroscopy. In: Theophile T editor. Infrared Spectroscopy-Life and Biomedical Sciences. InTech; 2012. Available from: http://www.intechopen.com/books/infrared-spectroscopy-life-and-biomedical-sciences/characterization-of-bone-and-bone-based-graft-materials-using-ftir-spectroscopy ISBN: 978-953-51-0538-1

[38] Kumar V, Vora SD, Asodiya FA, Kumar N, Gangwar AK. Fourier transform infrared spectroscopy of the animal tissues. In: Real Perspective of Fourier Transforms and Current Developments in Superconductivity. London, UK: IntechOpen; 2020. DOI: 10.5772/intechopen.94582

[39] Ren F, Ding Y, Leng Y. Infrared spectroscopic characterization of carbonated apatite: A combined experimental and computational study. Journal of Biomedical Materials Research Part A. 2014;**102A**:495-505

[40] Xu B, Poduska KM. Linking crystal structure with temperature-sensitive

vibrational modes in calcium carbonate minerals. Physical Chemistry Chemical Physics. 2014;**16**:17634. DOI: 10.1039/c4cp01772b

[41] Xu B, Toffolo MB, Regev L, Boaretto E, Poduska KM. Structural differences in archeological relevant calcite. Analytical Methods. 2015;7:9304. DOI: 10.1039/c5ay01942g

[42] Beasley MM, Baterlink L, Miller RM. Comparison of transmission FTIR, ATR, and DRIFT spectra: Implications for assessment of bone bioapatite diagenesis. Journal of Archaeological Science. 2014;**46**:16-22. DOI: 10.1016/j.jas.2014.03.008

[43] Beasley MM, Schoeninger MJ, Miller R, Bartelink EJ. Preservation of bone organic fraction is not predictive of the preservation of bone inorganic fraction when assessing stable isotope analysis sample quality control measures. Journal of Archaeological Science. 2024;**162**:105886. DOI: 10.1016/j.jas.2023.105886

[44] Caruso V, Marinoni N, Diella V, Possenti E, Mancini L, Cantaluppi M, et al. Diagenesis of juvenile skeletal remains: A multimodal and multiscale approach to examine the postmortem decay of children's bones. Journal of Archaeological Science. 2021;**135**:105477. DOI: 10.1016/j.jas.2021.105477

[45] Hollund HI, Ariese F, Fernandes R, Jans MME, Kars H. Testing an alternative high-throughput tool for investigating bone diagenesis: FTIR in attenuated total reflection (ATR) mode. Archaeometry. 2013;**55**:507-532. DOI: 10.1111/j.1475-4754.2012.00695.x

[46] Maurer AF, Zeitoun V, Bardin J, Millard AR, Ségalen L, Guérin F, et al. Multifactorial approach to describe early diagenesis of bones: The case study of the merovingian cemetery of saint-Linaire (France). Quaternary International. 2023;**660**:42-55. DOI: 10.1016/j.quaint.2023.03.003

[47] Mircea C, Vulpoi A, Rusu I, Radu C, Pârvu M, Kelemen B. Exploring post-excavation degradation potential of fungal communities associated with archaeological human remains. Archaeometry. 2019;**61**:750-763. DOI: 10.1111/arcm.12438

[48] Smith DR, Martin EK, Kaufman BL, Callaghan M, Cardona K, Kovacevich B, et al. The bottom line: Exploring analytical methods for assessing bioapatite preservation in archaeological bone using FTIR-ATR. Journal of Archaeological Science: Reports. 2023;**50**:104014. DOI: 10.1016/j.jasrep.2023.104014

[49] Colmenares-Prado M, Martínez CA, Veiga-Rilo C, López-Costas O. Application of FTIR spectroscopy to the study of ante- and post-morten changes in archeological human bones from a Lanzada necorpolis (NW Spain). Spectrochimica Acta. 2024 (under review)

[50] Hedges REM, Clement JG, Thomas CDL, O'Connell TC. Collagen turnover in the adult femoral mid-shaft: Modeled from anthropogenic radiocarbon tracer measurements. American Journal of Physical Anthropology. 2007;**133**:808-816

[51] Quinn RL. How much time is recorded with a rib bone isotope? Journal of Archaeological Science: Reports. 2024;**57**:104593

[52] López-Costas O, Lantes-Suárez Ó, Martínez CA. Chemical compositional changes in archaeological human bones due to diagenesis: Type of bone vs soil

environment. Journal of Archaeological Science. 2016;**67**:43-51

[53] Lambert JB, Vlasak SM, Thometz AC, Buikstra JE. A comparative study of the chemical analysis of ribs and femurs in woodland populations. American Journal of Physical Anthropology. 1982;**59**:289-294

[54] Zapata J, Perez-Sirvent C, Martinez-Sanchez MJ, Tovar P. Diagenesis, not biogenesis: Two late Roman skeletal examples. Science of the Total Environment. 2006;**369**:357-368

[55] Márquez-Grant N, Muñoz Villarreal AP, López-Costas O. Restos humanos subadultos: Cuestiones prácticas en antropología forense. In: Sanabria Medina C, editor. Antropología y patología forense de la muerte - La Investigación científico-judicial de la muerte y la tortura, desde las fosas clandestinas, hasta la audiencia pública. Bogotá: Forensic Publisher; 2016. pp. 26-60

Chapter 2

Approximation of Uniform Almost Periodic Functions by Inclusions and Integrals

Talbakov Farkhodjon Makhmadshoevich

Abstract

The paper examines some issues of approximation of almost periodic Bohr functions from partial sums of the Fourier series and Marcinkiewicz means, when the Fourier exponents (function spectrum) of the functions under consideration have a limit point at infinity. The question of the deviation of a given function $f(x)$ from its partial sums of the Fourier series is investigated, depending on the rate at which the value of the best approximation by a trigonometric polynomial of limited degree tends to zero. Here, when determining the Fourier coefficients, instead of the function under consideration, some arbitrary, real, continuous function is taken $\Phi_\sigma(t)$ $(\sigma > 0)$, which in a given interval is equal to one, and in other cases is equal to zero. Next, an upper estimate for the deviation of an almost periodic function in the sense of Bohr by Marcinkevich means is established in a similar manner.

Keywords: uniform almost periodic functions, Fourier series, spectrum of a function, Fourier coefficients, limit point at infinity, Marcinkiewicz means, trigonometric polynomial, best approximation

1. Introduction

Let $L_p(1 \leq p < \infty)$ be the space of measurable functions $f(x,y)$, of period 2π for each variable, with norm

$$\|f\|_{L_p} = \left\{ \int_0^{2\pi} \int_0^{2\pi} |f(x,y)|^p dxdy \right\}^{\frac{1}{p}} < \infty,$$

and when $p = \infty$

$$\|f\|_{L_p} = \underset{x,y}{vraisup} |f(x,y)| < \infty.$$

The Fourier series of this function has the form

$$f(x,y) \sim \sum_{k=0}^{\infty}\sum_{l=0}^{\infty} A_{k,l}(x,y),$$

where the coefficients of this series are determined by the formulas

$$A_{k,l}(x,y) = a_{k,l}\cos kx\cos ly + b_{k,l}\sin kx\cos ly$$
$$+c_{k,m}\cos kx\sin ly + d_{k,m}\sin kx\sin ly(k,l \leq 1),$$
$$A_{k,0}(x,y) = \frac{1}{2}(a_{k,0}\cos kx + b_{k,0}\sin kx\ ),$$
$$A_{0,l}(x,y) = \frac{1}{2}(a_{0,l}\cos ly + c_{0,l}\sin ly),$$
$$A_{0,0} = \frac{1}{4}a_{0,0}.$$

In 1964, Martsinkevich [1] investigated the issue of deviation of sums of the form

$$\sigma_n(f;x,y) = (n+1)^{-1}\sum_{k=0}^{m} S_{k,k}(f;x,y),$$

where $S_{k,k}(f;x,y)$ is the partial sum of order k for each of the variables of the Fourier series of the function $f(x,y) \in L_p\ (1 \leq p \leq \infty)$. In particular, he pointed out that if $f(x,y)$ is continuous over the set of variables x, y, then the estimate holds

$$R_n(f)_{L_\infty} = \|f(x,y) - \sigma_n(f;x,y)\|_{L_\infty} \to 0\ (n \to \infty).$$

In the work of Zhizhiashvili [2], some estimates were obtained about the rate at which the quantity tends to zero

$$R_n(f)_{L_p} = \|f(x,y) - \sigma_n(f;x,y)\|_{L_p}\ (1 < p \leq \infty).$$

Behavior of the deviation of a function of two variables $f(x,y) \in L_p\ (1 \leq p \leq \infty)$ from sums of the form

$$W_r(f;x,y) = (1-r)\sum_{k=0}^{\infty} r^k S_{k,k}(f;x,y)$$

for $0 < r < 1$ and $r \to \infty$ and were studied in the work of Tabersky ([3] p. 310). Some clarifying assessments of the results of Zhizhiashvili and Tabersky are given in the work of Timan and Gaimnazarov [4] in 1972.

Timan and Ponomarenko [5] investigated the question of the behavior of the quantity

$$R_n(f;\nu)_{L_p} = \left\| f(x,y) - \sum_{k=0}^{n} \nu_{k,k}(f;x,y) \right\|_{L_p} \tag{1}$$

for classical triangular matrices $\{\mu_{k,n}\}$

$$\{\mu_{k,n}\} = \begin{cases} \dfrac{A_{n-k}^{\alpha-1}}{A_k^{\alpha}}, k \leq n; \\ 0, k > n. \end{cases}$$

For quantity (1), an upper bound is given in [6], and a lower bound for this quantity is established in [4]. In [5], some estimates from the works of L.V. Zhizhiashvili [2] and R. Taberski [3] were also refined. And also for any natural r in [7] exact order equalities for a triangular matrix were obtained

$$\{\mu_{k,n}\} = \begin{cases} \dfrac{(k+1)^r - r^k}{(n+1)^r}, k \leq n; \\ 0, k > n. \end{cases} \tag{2}$$

It has been established that if $f(x,y) \in L_p$ and $\{\mu_{k,n}\}$are defined in the form of a triangular matrix (2), then the inequality

$$R_n(f;\nu)_{L_p} \leq c_r(n+1)^{-r} \sum_{\nu=0}^{n} (\nu+1)^{r-1} E_{\nu,\nu}(f)_{L_p},$$

Where

$$E_{\nu,\nu}(f)_{L_p} = \inf_{T_{k,k}} \|f(x,y) - T_{\nu,\nu}(x,y)\|_{L_p},$$

$T_{\nu,\nu}$ is the trigonometric polynomials of order not higher than n in each of the variables, and the constant c_r depends only on r.

If $f(x,y) \in L_p$ $(1 < p \leq \infty)$ and $\{\mu_{k,n}\}$ are defined by relation (2), then in ([2], pp. 549–550), estimates were obtained using continuity modules, that is,

$$R_n(f;\nu)_{L_p} \leq c_{p,k}\left(\omega_k^{(1)}\left(f;\frac{1}{n}\right)_{L_p} + \omega_k^{(2)}(f;\frac{1}{n})_{L_p}\right),$$

Where

$$\omega_k^{(1)}(f;u)_{L_p} = \sup_{|h| \leq u} \|\Delta_{x,h}^k f\|_{L_p}, \omega_k^{(2)}(f;u)_{L_p} = \sup_{|h| \leq u} \left\|\Delta_{h,y}^k f\right\|_{L_p},$$

$$\Delta_{x,h}^k f = \sum_{\nu=0}^{k} (-1)^{k-r} \binom{k}{\nu} f(x + \nu h, y),$$

$$\Delta_{h,y}^k f = \sum_{\nu=0}^{k} (-1)^{k-r} \binom{k}{\nu} f(x, y + \nu h)(u > 0, k \in N),$$

$c_{p,k}$ is a constant depending on p and k.

For $1 < p < \infty$ in [5], the following results were also obtained:

a. $\omega_k^{(1)} \leq M_{p,k} R_n(f;\mu)_{L_p}, \omega_k^{(2)} \leq M_{p,k} R_n(f;\mu)_{L_p},$

where $M_{p,k}$ is a constant depending on p and k;

b. $R_n(f;\mu)_{L_p} \leq \omega_k^{(1)}\left(f;\frac{1}{n}\right)_{l_p} + \omega_k^{(2)}\left(f;\frac{1}{n}\right)_{L_p}$;

c. $\|f(x,y) - W_r(f;x,y)\|_{L_p} \leq c_p\left(\omega_1^{(1)}(f;1-r)_{L_p} + \omega_1^{(2)}(f;1-r)_{L_p}\right)$;

d. $\omega_1^{(1)}(f;1-r)_{L_p} \leq M_p\|f(x,y) - W_k(f;x,y)\|_{L_p}$;

e. $\omega_1^{(2)}(f;1-r)_{L_p} \leq M_p\|f(x,y) - W_k(f;x,y)\|_{L_p}$.

In the work of Khasanova [7], they obtained analogues of the above results on the behavior of deviations of functions of two variables $f(x,y)$, given on the plane from the integral means of their Fourier transforms in the metric of the space $L_p(R^2)$ $(1 \leq p < \infty)$.

Let $f(x,y) \in L_p(R^2)$ be the space of measurable functions $f(x,y)$ for which

$$\|f\|_{L_p} = \left\{\int_{-\infty}^{\infty}\int_{-\infty}^{\infty} |f(x,y)|^p dxdy\right\}^{\frac{1}{p}} < \infty,$$

and almost everywhere, there is a Fourier transform

$$F(t,z) = \frac{1}{2\pi}\int_{-\infty}^{\infty}\int_{-\infty}^{\infty} f(u,v)\exp(-i(tu+zv))dudv,$$

where $F(t,z) \in l_q(R^2)$ $\left(\frac{1}{p}+\frac{1}{q}=1\right)$.

For $\sigma > 0$, consider the partial sums

$$\begin{aligned} S_{\sigma,\sigma}(f;x,y) &= \int_{-\sigma}^{\sigma}\int_{-\sigma}^{\sigma} F(t,z)e^{i(tx+zy)}dtdz \\ &= \int_{-\sigma}^{\sigma}\left[\int_{-u}^{u} A(t,u)dt + \int_{-u}^{u} A(t,-u)dt + \int_{-u}^{u} A(u,z)dz + \int_{-u}^{u} A(-u,z)dz\right] \\ &= \int_{0}^{\sigma} S^*_{u,u}(f;x,y)du, \end{aligned} \tag{3}$$

where

$$A(t,z) = F(t,z)e^{i(tx+zy)},$$

$$S^*_{u,u}(f;x,y) = \int_{-u}^{u} A(t,u)dt + \int_{-u}^{u} A(t,-u)dt + \int_{-u}^{u} A(u,z)dz + \int_{-u}^{u} A(-u,z)dz.$$

We investigate the order of behavior of the deviation value

$$R_{\sigma,r}(f)_{L_p} = \|f(x,y) - U_{\sigma,r}(f;x,y)\|_{L_p}, \tag{4}$$

where

$$U_{\sigma,r}f(x,y) = \int_0^{\sigma}\left(1-\frac{u^r}{\sigma^r}\right)S^*_{u,u}(f;x,y).$$

The following statements hold [7]:

Theorem A. If $f \in L_p(R^2)$ $(1 < p \le 2)$, then the estimate holds

$$R_{\sigma,k}(f;)_{L_p} \le c_{p,k}\left(\omega_k^{(1)}\left(f;\frac{1}{\sigma}\right)_{L_p} + \omega_k^{(2)}(f;\frac{1}{\sigma})_{L_p}\right),$$

where $c_{p,k}$ is a constant that depends only on p and k.

Theorem B. Let $f(x,y) \in L_p(R^2)$ $(0 < p \le 2)$. Then, it holds

$$\omega_k^{(\nu)}\left(f;\frac{1}{\sigma}\right)_{L_p} \le M_{p,k}R_{\sigma,k}(f)_{L_p}\ (\nu = 1, 2),$$

$M_{p,k}$ is a constant that depends only on p and k.

From Theorems A and B, it follows that if the conditions of Theorem B hold, then for any natural k, the estimate

$$R_{\sigma,k}(f;)_{L_p} \le c_{p,k}\left(\omega_k^{(1)}\left(f;\frac{1}{\sigma}\right)_{L_p} + \omega_k^{(2)}(f;\frac{1}{\sigma})_{L_p}\right).$$

For $p = \infty$ and $k = 1$, similar results were obtained in [9] and for the case of periodic functions in [4].

2. Main results

Now we present the results of this section for estimating from above the deviation of functions $\boldsymbol{f}(\boldsymbol{x}) \in \boldsymbol{B}$ from Marcinkiewicz-type sums when the Fourier exponents have a single limit point at infinity.

Let $B(R)$ be the space of all bounded functions $f(x) \in \boldsymbol{B}$ with norm

$$\|f\|_{\boldsymbol{B}} = \sup_{x\in R}|f(x)|$$

and the function $f(x) \in \boldsymbol{B}$ has a Fourier series of the form

$$f(x) \sim \sum_{k=-\infty}^{\infty} A_k \exp(i\lambda_k x), \tag{5}$$

with odds

$$A_k = \lim_{T\to\infty}\frac{1}{2T}\int_{-T}^{T} f(x)\exp(-i\lambda_k x)dx,$$

where numbers $\{\lambda_k\}$ are Fourier exponents that have a unique limit point at infinity, that is (see, for example, ([10], p. 814) and ([11], p. 55)),

$$\lambda_0 = 0, \lambda_{-n} = -\lambda_n, |\lambda_n| < |\lambda_{n+1}|, \lim_{n\to\infty} \lambda_n = \infty.$$

Let us introduce into consideration

$$S_\sigma(f;x)_B = \sum_{|\lambda_n|\leq\sigma} A_n \Phi_\sigma(\lambda_n) e^{i\lambda_n x},$$

where $\Phi_\sigma(t) \in R$ is a continuous and even function such that $\Phi_\sigma(0) = 1$, $\Phi_\sigma(t) = 0$ at $|t| \leq \sigma$;

$$\psi_\sigma(t) \in L(-\infty, \infty), \tag{6}$$

where

$$\psi_\sigma(u) = \frac{1}{2\pi} \int_{-\infty}^{\infty} \Phi_\sigma(t) e^{-iut} dt.$$

First, we prove the following auxiliary statement, which has an independent character.

Lemma 1 [12]. *If the function* $f(x) \in \boldsymbol{B}$, *then*

$$S_\sigma(f;x) = U_\sigma(f;\varphi;x) = \int_{-\infty}^{\infty} f(x+t)\Phi_\sigma(t)dt.$$

Proof. Let

$$f_\sigma(x) = \int_{-\infty}^{\infty} f(x+t)\Phi_\sigma(t)dt$$

Then, thanks to (6), we obtain that

$$|f_\sigma(x+\tau) - f_\sigma(x)| \leq \int_{-\infty}^{\infty} |f(x+t+\tau) - f(x+t)||\psi_\sigma(t)|dt \leq \sup_x |f(x+\tau) - f(x)| \int_{-\infty}^{\infty} |\psi_\sigma(t)|dt.$$

The integral on the right side of the last inequality converges and

$$|f(x+t+\tau) - f(x+t)| < \varepsilon.$$

Therefore, from the last inequality, it follows that the function $f(x) \in \boldsymbol{B}$.

Let $D(\lambda_n)$ denote the Fourier coefficients of the functions $f_\sigma(x) \in \boldsymbol{B}$, which correspond to the exponents λ_n. Then, we have

$$\frac{1}{2T}\int_{-T}^{T} f_\sigma(x)e^{-i\lambda_n x}dx = \frac{1}{2T}\int_{-T}^{T}\int_{-\infty}^{\infty} f(x+t)\psi_\sigma(t)e^{-i\lambda_n x}dtdx$$
$$= \int_{-\infty}^{\infty}\psi_\sigma(t)dt\frac{1}{2T}\int_{-T}^{T} f(x+t)e^{-i\lambda_n x}dx$$
$$= \int_{-\infty}^{\infty}\psi_\sigma(t)\left(\frac{1}{2T}\int_{-T}^{T} f(x+t)e^{-i\lambda_n x}dx\right)dt$$
$$= \int_{-\infty}^{\infty}\psi_\sigma(t)\left(\frac{1}{2T}\int_{-T+t}^{T+t} f(x)e^{-i\lambda_n (x+t)}dx\right)dt$$
$$+ \int_{-\infty}^{\infty}\psi_\sigma(t)\left(\frac{1}{2T}\int_{-T+t}^{T+t} f(x)e^{-i\lambda_n x}e^{-i\lambda_n t}dx\right)dt$$
$$= \int_{-\infty}^{\infty}\psi_\sigma(t)e^{-i\lambda_n t}\left(\frac{1}{2T}\int_{-T+t}^{T+t} f(x)e^{-i\lambda_n x}dx\right)dt.$$

The internal integral in the latter is the pre-limit expression of the Fourier coefficients of the function $f(x)$. So, according to the Fourier inversion formula

$$\int_{-\infty}^{\infty}\psi_\sigma(t)e^{-i\lambda_n t}dt = \Phi_\sigma(\lambda_n).$$

Therefore, $D(\lambda_n) = A_n\Phi_\sigma(\lambda_n)$ and

$$f_\sigma(x) = \sum_{|\lambda_n|\le\sigma} A_n\Phi_\sigma(\lambda_n)e^{i\lambda_n x}.$$

Since $f(x)\in\boldsymbol{B}$, the statement of the lemma follows from the latter.
Let $B(R)$ be the space of all bounded functions $f(x)\in\boldsymbol{B}$ with norm

$$\|f\|_{\boldsymbol{B}} = \sup_{x\in R}|f(x)|.$$

Let us consider the quantity

$$R(f;x) = \|U_\sigma(f;\varphi;x) - f(x)\|_{\boldsymbol{B}}, \tag{7}$$

Wherein

$$U_\sigma(f;\varphi;x) = \int_{-\infty}^{\infty} f(x+t)\Phi_\sigma(t)dt,\ \Phi_\sigma(t) = \frac{1}{2\pi}\int_0^{\infty}\varphi_\sigma(u)K_m(t)du, \tag{8}$$

where $K_m(t) = \frac{4\sin(ut)}{t}$,$\varphi_\sigma(u)$ is some absolutely integrable on the interval $(0, \infty)$ function for every fixed $\sigma > 0$ and such that

$$\int_{-\infty}^{\infty} |\Phi_\sigma(t)|dt = 1. \tag{9}$$

Depending on the speed of approaching zero $E_\sigma(f)$ $(\sigma \to \infty)$ for cases when

$$\varphi_\sigma(u) = \varphi_{\sigma,a}(u) = \begin{cases} 1, |u| \leq a \ (0 < a < \sigma); \\ \dfrac{\sigma - |u|}{\sigma - a}, a < |u| < \sigma; \\ 0, |u| \geq \sigma, \end{cases} \tag{10}$$

Let us study the behavior of quantity (7).

It is known that [5] (see [5], p. 60, Theorem 1)

$$\int_{-\infty}^{\infty} |\Phi_{\sigma,a}(t)|dt \leq C\frac{\sigma + a}{\sigma - a}, \tag{11}$$

where C is a constant.

Theorem 1 [12]. *Let $f(x) \in B(R)$ and the function $\varphi_\sigma(u) = \varphi_{\sigma,a}(u)$ be defined by the relation (10). Then, for any Λ $(0 < \Lambda < a < \sigma)$, the following estimate holds:*

$$R(f;\varphi_\sigma) \leq C\frac{\sigma + a}{\sigma - a}E_\Lambda(f)_B, \tag{12}$$

where C is an absolute constant.

Proof. Let us write relation (9) in the form

$$\int_0^\infty \Phi_\sigma(t)dt = \frac{1}{2}.$$

Let us multiply both sides of the latter by $f(x)$ and subtract the resulting equality from (8) with $\Phi_\sigma(t) = \Phi_{\sigma,a}(t)$, we will have

$$\begin{aligned} \Delta_{\sigma,a}(f;x) &= U_\sigma(f;\varphi;x) - f(x) \\ &= \int_{-\infty}^{\infty} (f(x+t)\Phi_{\sigma,a}(t) - f(x)\Phi_{\sigma,a}(t))dt \\ &= \int_0^\infty [f(x+t) - f(x-t)]\Phi_{\sigma,a}(t)dt - 2\int_0^\infty f(x)\Phi_{\sigma,a}(t)dt = \\ &= \int_0^\infty [f(x+t) - f(x-t) - 2f(x)]\Phi_{\sigma,a}(t)dt \\ &= \int_0^\infty \Omega_x(f;t)\Phi_{\sigma,a}(t)dt, \end{aligned}$$

where $\Omega_x(f;t) = f(x+t) + f(x-t) - 2f(x)$.

Let

$T_r = \sum_{|\lambda_m| \le r} A_m e^{i\lambda_m x}$ be the

arbitrary trigonometric polynomial and $0 < \Lambda < a < \sigma$. Then (see ([12], p. 14), Theorem 3),

$$T_\Lambda(x) = \int_{-\infty}^{\infty} \sum_{|\lambda_m| \le r} T_\Lambda f(x+t)\Phi_{\sigma,a}(t).$$

It is easy to show that, for the polynomial $T_\Lambda(x)$,

$$\int_0^\infty \Omega_x(T_\Lambda; f; t)\Phi_{\sigma,a} dt = 0.$$

Indeed, by virtue of estimate (9), we obtain

$$\begin{aligned}
&\int_0^\infty \Omega_x(T_\Lambda; f; t)\Phi_{\sigma,a}(t)dt \\
&= \int_0^\infty [T_\Lambda(x+t) + T_\Lambda(x-t) - 2T_\Lambda(x)]\Phi_{\sigma,a}(t)dt \\
&= \int_{-\infty}^{\infty} T_\Lambda(x+t)\Phi_{\sigma,a}(t)dt - \int_{-\infty}^{\infty} T_\Lambda(x)\Phi_{\sigma,a}(t)dt \\
&= T_\Lambda(x) - T_\Lambda(x)\int_{-\infty}^{\infty} \Phi_{\sigma,a}(t)dt = 0
\end{aligned}$$

Means

$$\Delta_{\sigma,a}(f;x) = \int_0^\infty \Omega_x[(f - T_\Lambda);t]\Phi_{\sigma,a}(t)dt. \tag{13}$$

Let

$$\|f(x) - T_\Lambda(x)\| = E_\Lambda(f)_B,$$

where $T_r(x)$ is a trigonometric polynomial that provides the best approximation of order Λ, then

$$\|\Omega_x[(f - T_\Lambda);t]\|_B = 4E_r(f)_B. \tag{14}$$

From (11), (14), and (13), estimate (12) follows. Theorem 1 is proven.

Theorem 2 [12]. *Let* $f(x) \in \boldsymbol{B}$. *Then, the estimate is valid*

$$\left\| f(x) - (n+1)^{-1} \sum_{k=0}^{n} S_k(f;x) \right\|_B \le M(n+1)^{-1} \sum_{k=0}^{n} E_\Lambda(f)_B, (Б)$$

where M is a constant and

$$E_\Lambda(f;x)_B = \inf_{A(\lambda_n)} \left\| f(x) - \sum_{|\lambda_n| \le k} A_n e^{i\lambda_n x} \right\|_B,$$

the value of the best approximation of the function $f(x)$ by trigonometric polynomials of degree not higher than Λ.

Proof. Let $n \in [2^m; 2^m + 1]$. Then,

$$R_n(f)_B = \left\| f(x) - (n+1)^{-1} \sum_{k=0}^{n} S_k(f;x) \right\|_B$$

$$= \left\| (n+1)^{-1} \sum_{k=0}^{n} (f(x) - S_k(f;x)) \right\|_B$$

$$= (n+1)^{-1} \left\| \sum_{k=0}^{m-1} \sum_{k=2^\nu}^{2^{\nu+1}-1} (f(x) - S_k(f;x)) + f(x) - S_0(f;x) + \sum_{r=2^m}^{n} (f(x) - S_k(f;x)) \right\|_B$$

$$\le (n+1)^{-1} \left\| \sum_{\nu=0}^{n-1} 2^\nu \bullet \frac{1}{2^\nu} \sum_{k=2^\nu}^{2^{\nu+1}-1} (f(x) - S_k(f;x)) \right\|_B$$

$$+ (n+1)^{-1} \| f(x) - S_0(f;x) \|_B + (n+1)^{-1} \left\| \sum_{k=2^m}^{n} (f(x) - S_k(f;x)) \right\|_B. \tag{15}$$

By Theorem 1, we have

$$\left\| 2^{-\nu} \sum_{k=2^\nu}^{2^{\nu+1}-1} (f(x) - S_k(f;x)) \right\|_B \le M E_{2^{\nu+1}-1}(f)_B \tag{16}$$

$$\left\| \sum_{k=2^\nu}^{m} (f(x) - S_k(f;x)) \right\|_B \le M(n - 2^m) E_{2^m-1}(f)_B. \tag{17}$$

Substituting (16) and (17) into (15), we obtain

$$R_n(f)_B \le M(n+1)^{-1} \sum_{\nu=0}^{m-1} 2^\nu E_{2^n-1}(f)_B + (n+1)^{-1} E_0(f)_B$$

$$+ M(n+1)^{-1}(n - 2^m) E_{2^m-1}(f)_B$$

$$\le M(n+1)^{-1} \sum_{\nu=0}^{m-1} 2^\nu E_{2^\nu-1}(f)_B + (n+1)^{-1} E_0(f)_B$$

$$+ M(n+1)^{-1} E_{2^m-1}(f)_B \le M_1(n+1)^{-1} \sum_{k=0}^{2^m} E_k(f)_B \le M_1(n+1)^{-1} \sum_{k=0}^{n} E_k(f)_B.$$

This implies inequality (B), which completes the proof of Theorem 2.

An analogue of Theorem 2 for periodic functions was established in the work of Timan and Ponomarenko [5], and for the class of almost periodic functions in the sense of Bohr and Besicovitch in [6, 9–17].

Let $\boldsymbol{B}_\sigma$ $(\sigma > 0)$ denote the class of entire functions $g_\sigma(x)$ of degree not higher than σ that are bounded on the entire real axis. Bernstein [7] established that, among the functions from the class $\boldsymbol{B}_\sigma$ that provide the best approximation of the 2π-periodic function $f(x)$ on the entire real axis, there is a trigonometric polynomial of degree no higher than σ.

Analogs of the theorem of S.N. Bernstein for functions $f(x) \in \boldsymbol{B}$ were obtained in the works of Bredikhina [8, 13]. She proved that if a function $f(x) \in \boldsymbol{B}$ with Fourier series

$$f(x) \sim \sum_k A_k e^{i\lambda_k \beta x},$$

where λ_k are rational numbers and β is a real number, then among the functions $g_\sigma(x) \in \boldsymbol{B}_\sigma$ $(\sigma > 0)$, for which

$$\sup_{-\infty < x < \infty} |f(x) - g_\sigma(x)| = A_\sigma(f), \tag{18}$$

there is a trigonometric polynomial of degree ≤σ.

The space of uniform almost-periodic functions, denoted by $\boldsymbol{B}$, is the closure of the set of trigonometric polynomials

$$T_k(x) = \sum_{k=1}^{n} A_k e^{i\lambda_k x},$$

where A_k are the Fourier coefficients, λ_k is the spectrum of the function $f(x) \in \boldsymbol{B}$, with norm

$$\|f(x)\|_B = \sup_{x \in R} |f(x)|.$$

Let a function $f(x) \in \boldsymbol{B}$ with an arbitrary spectrum $\{\lambda_k\}$, have a Fourevied series

$$f(x) \sim \sum_{k=-\infty}^{\infty} A_k \mathrm{e}^{i\lambda_k x}, \tag{19}$$

where the Fourier coefficients $\{A_k\}$ are defined as follows

$$A_k = M\{f(x)\mathrm{e}^{-i\lambda_k x}\} = \lim_{T \to \infty} \frac{1}{2T} \int_{-T}^{T} f(x)\mathrm{e}^{-i\lambda_k x} dx.$$

Consider the following problem. Let the function $f(x) \in \boldsymbol{B}$. What are the necessary and sufficient conditions for this function to belong to the class $\boldsymbol{B}_\sigma$. The answer to this question is given as follows.

Theorem 3. In order for $f(x) \in \boldsymbol{B}$ to belong to the class of entire functions $\boldsymbol{B}_\sigma$, it is necessary and sufficient that the condition

$$|\lambda_k| \leq \sigma. \tag{20}$$

Proof. Adequacy. Consider the function

$$f_{a,b}(x) = \int_{-\infty}^{\infty} f(x+t)\varphi_{a,b}(t)dt,$$

where

$$\varphi_{a,b}(t) = \frac{2}{\pi(a-b)t^2} \bullet sin\,\frac{a-b}{2}tsin\frac{b+a}{2}t.$$

It is easy to show that the function $f_{a,b}(x)$ is continuous and almost periodic. Indeed, since [12] (see [12], p. 74)

$$\int_{-\infty}^{\infty} |\varphi_{a,b}(t)|dt \leq A + Bln\frac{a+b}{b-a},$$

where A and B are constants, then

$$|f_{a,b}(x+\tau) - f_{a,b}(x)| \leq \int_{-\infty}^{\infty} |f(x+t+\tau) - f(x+t)| \bullet |\varphi_{a,b}(t)|dt \leq$$
$$\leq \sup_{-\infty<x<\infty} |f(x+t+\tau) - f(x+t)| \left(A + Bln\frac{a+b}{b-a}\right).$$

This implies continuity and almost periodicity of the function $f_{a,b}(x) \in \mathbf{B}$. Let us put $a = \lambda, b = 2\lambda$. Then,

$$f_{a,b}(x) = \frac{2}{\lambda\pi}\int_{-\infty}^{\infty} f(t)\frac{sin\,\frac{(t-x)\lambda}{2}\,sin\,\frac{3\lambda(t-x)}{2}}{(t-x)^2}dt,$$

or

$$f_{a,b}(x) \sim \sum_{k=-\infty}^{\infty} A_k \exp(i\lambda_k x).$$

Hence, thanks to the uniqueness theorem [12] (see [12], pp. 65–67), it will be

$$f_{a,b}(x) \to f(x).$$

Necessity. Let the function $f(x) \in \mathbf{B}$ belong to the class $\mathbf{B}_\sigma$ and have a Fourier series of the form (19). Then, for any natural r, the derivative of order r will be

$$f^{(r)}(x) \sim \sum_{k=-\infty}^{\infty} (i\lambda_k)^r A_k\, exp\,(i\lambda_x x).$$

Using inequality S.N. Bernstein, we come to

$$\left|f^{(r)}(x)\right| \le \sigma^r \bullet \sup_x |f(x)| = \sigma^r \bullet C.$$

It means

$$\lim_{T\to\infty} \frac{1}{2T} \int_{-T}^{T} \left|f^{(r)}(x)\right|^2 dx \le \sigma^{2r} \bullet C^2.$$

From here, applying Bessel's inequalities, we have

$$|\lambda_k|^{2r} |A_k|^2 \le \sigma^{2r} \bullet C^2 (r = 1, 2, \dots),$$

or

$$|A_k| \le C \bullet \left(\frac{\sigma}{|\lambda_k|}\right)^r.$$

Therefore, for $|\lambda_k| > \sigma$ coefficients $A_k = 0$. Theorem 1 is proven.

It is known [10] that, from the set of functions $f(x) \in \boldsymbol{B}_\sigma$, which provide the best approximation of the functions $f(x) \in L_p[-\pi, \pi]$ on the entire real axis, one can find a trigonometric polynomial of degree $\le \sigma$.

This statement is based on the fact that if $g_\sigma(f; x) \in \boldsymbol{B}_\sigma$

$$\sup_{x\in R} |f(x) - g_\sigma(f; x)| = A_\sigma(f),$$

then the relations are valid uniformly over all $x \in R$

$$\left| f(x) - \frac{1}{2n+1} \sum_{k=-n}^{n} g_\sigma(x + 2k\pi) \right| \le A_\sigma(f), \tag{21}$$

$$\lim_{n\to\infty} \frac{1}{2n+1} \sum_{k=-n}^{n} \{g_\sigma(x + 2k\pi + 2\pi) - g_\sigma(x + 2k\pi)\} = 0, \tag{22}$$

where $A_\sigma(f)$ is the best approximation of order σ.

This result can be obtained if (21) and (22) are replaced by the relations

$$|\Phi_n(x) - Q_{\sigma,N,n}(x)| \le A_\sigma(f),$$

where $\sigma > 0; N, n$ are any natural numbers,

$$\Phi_n(x) = \frac{1}{\pi} \int_{-\pi}^{\pi} f(x+t) F_n(t) dt = \frac{1}{2\pi N} \int_{-2\pi N}^{2\pi N} f(x+t) F_n(t) dt,$$

$$Q_{\sigma,N,n}(x) = \frac{1}{2\pi N} \int_{-2\pi N}^{2\pi N} g_\sigma(x+t) F_n(t) dt,$$

$$F_n(t) = \frac{\sin^2(n+1)\frac{t}{2}}{2(n+1)\sin^2\frac{t}{2}},$$

and for all fixed n uniformly in $x \in R$

$$\lim_{N\to\infty}\{Q_{\sigma,N,n}(x+2\pi) - Q_{\sigma,N,n}(x)\} = 0.$$

When establishing similar results for functions $f(x) \in \boldsymbol{B}$, a number of problems arise. Firstly, additional conditions are imposed on the smoothness of functions, and secondly, the Fourier exponents of such functions can lie densely everywhere, that is, [6, 11, 12, 15–17]:

$$\lambda_0 = 0; \lambda_{-k} = -\lambda_k, |\lambda_k| > |\lambda_{k+1}|, (k = 1, 2, \ldots), \lim_{k\to\infty}|\lambda_k| = 0;$$

$$\lambda_0 = 0; \lambda_{-k} = -\lambda_k, |\lambda_k| < |\lambda_{k+1}|, (k = 1, 2, \ldots), \lim_{k\to\infty}|\lambda_k| = \infty.$$

Author details

Talbakov Farkhodjon Makhmadshoevich
Department of Geometry and Higher Mathematics, Tajik State Pedagogical University named after S. Aini, Dushanbe, Tajikistan

*Address all correspondence to: talbakov_90@mail.ru

References

[1] Marcinkewisz I. Sur Une Metode Remarquable de Sommation des Series Doubles de Fourier: Collected Papers. Warszawa; 1964. pp. 527-538

[2] Zizhiashvili LV. On summation of double Fourier series L.V Zizhiashlivi. Siberian Mathematical Journal. 1967;**VIII** (3):548-564

[3] Taberski R. Abel summability of double Fourier series. Bull. Acad. Polon. Sci. Ser. Sci. Msth. Astron. Et Phys. 1970; **18**(6):307-314

[4] Timan MF, Gaimnazarov G. Evasion of periodic functions of two variables from some polynomials. Reports of the Academy of Sciences of the Taj SSR. 1972;**15**(5):6-8

[5] Timan MF, Ponomarenko VG. On the approximation of functions of two variables by sums of the Martsinkevich type. Izvestiya vuzov. Mathematics. 1975;**9**:59-67

[6] Talbakov FM. On the absolute convergence of double Fourier series of almost-periodic functions in a uniform metric. Russian Mathematical Survey. 2023;**67**(4):55-65. DOI: 10.3103/S1066369X2304006014

[7] Bernstein SN. Collected Works. Vol. 2. Moscow: Publishing House of the USSR Academy of Sciences; 1954. p. 627

[8] Bredikhina EA. On the issue of approximation of almost periodic functions. Sib. Math. Magazine. 1964;**5** (4):768-773

[9] Bohr G. Almost Periodic Functions. Gos. Teor. Izd-vo, Moscow: LIBROKOM; 2009. p. 128

[10] Timan AF. The Theory of Approximation of Functions of a Real Variable. Fizmatgiz; 1960. p. 624

[11] Khasanov YK, Talbakov FM. On the absolute convergence of double Fourier series of almost-periodic functions. Russian Mathematical Survey. 2024;**68** (4):60-71. DOI: 10.3103/S1066369X24700282

[12] Talbakov FM. On the Absolute Convergence of Fourier Series of Uniform Almost-periodic Functions and Some Questions of their Approximation. Juraev: Dissertation for the degree of Candidate of Physical and Mathematical Sciences NANT Institute of Mathematics named after A; 2020. p. 84

[13] Bredikhina EA. To the theorem of S. N. Bernshtein on the best approximation of continuous functions by entire functions of a given degree. Izv. Universities Mathematics. 1961;**6**:3-7

[14] Levitan BM. Almost Periodic Functions. Gos. Teor. Izd-vo, Moscow: GITTLE; 1953. p. 396

[15] Talbakov FM. On the absolute convergence of Fourier series of almost-periodic functions in uniform metrics 593 DAN RT. 2020;**63**(5-6):288-292

[16] Khasanov YK, Talbakov FM. On Sufficient Condition for Absolute Convergence Fourier Series of Bezikovichs Almost-Periodic Functions. Lobachevskii Journal of Mathematics. 2024;**45**(6):2743-2752. DOI: 10.1134/S199508022460290X. ISSN 1995-0802

[17] Talbakov FM. About Absolute Convergence of Fourier Series of Almost Periodic Functions. Pure and Applied Mathematics Journal. 2024;**13**(3):36-43. DOI: 10.11648/j.pamj.202420241303.11

Chapter 3

Modeling Dynamic Relationships between Energy Prices and Inflation in Euro Area Using Wavelets

Huthaifa Sameeh Alqaralleh and Alessandra Canepa

Abstract

In this chapter, we investigate the relationship between energy prices, economic uncertainty, and inflation in Eurozone countries. To account for cyclicality in the business cycle, the level of industrial production is also considered. We propose a wavelet time-varying vector autoregression model (VAR) model to assess the transmission of shocks across different time scales. The main findings of this analysis can be summarized as follows. First, a significant correlation is found between energy prices and inflation. Statistically significant periods of high correlation are detected primarily during times of turmoil, such as the 9/11 terrorist attacks, the U.S.-led invasion of Iraq in 2003, the global financial crisis, and the sovereign debt and banking crises. Second, the impact of uncertainty shocks on energy markets appears to be persistent. Third, energy prices serve as important shock transmitters, although the magnitude of this transmission varies over time and across different frequencies. Specifically, our estimation results suggest that energy price shocks are transmitted to both industrial production and inflation, with the impact being most noticeable around 8 months after the initial shocks. These results imply that while there is a pass-through effect, the transmission of energy price shocks to the economic system in the Eurozone is relatively slow.

Keywords: energy prices, inflation, economic uncertainty, wavelet time-varying vector autoregression, wavelet coherence analysis

1. Introduction

Since the seminal work by Hamilton [1], in the literature, a great deal of theoretical and empirical works has supported the view that energy price shocks, especially oil price fluctuations, have negative impact on output and inflation (see [2–10).

Against this background, the aim of this chapter is to investigate the dynamic relationships between energy prices and inflation in the euro area. The euro area case is of particular interest since despite being is one of the most industrialized economic regions in the world, energy production is largely undersized compared with the amount of energy required for final consumption [11]. The European countries are heavily dependent on crude oil imports, and the demand for natural gas has steadily increased in the last 50 years. This high level of energy dependency leaves the euro area exposed to energy price shocks. Accordingly, questions we address in this chapter are: How do shocks to energy prices affect inflation and output in the euro area countries? Also, the Covid-19 and the Ukraine conflict have highlighted the importance of accounting for economic uncertainty and geopolitical events when modeling the relationship between energy prices and inflation. A related question is: To what extent geopolitical events and economic uncertainty affect energy prices and inflation?

With respect to the existing literature, we take a fresh look at the modeling issue and apply the wavelet methodology to examine the relationships between the time series under consideration in both the time and frequency domain. In the literature, most of the related empirical works use linear or nonlinear models in conventional time-domain framework to investigate the relationship between energy price and inflation (e.g., [3, 11–17]).

In this chapter, we argue that existing studies on the relationship between energy prices and inflation may suffer from two limitations. First, a possible shortcoming of time-domain specifications is that these models offer limited insights into the time-varying features of the time series at hand. In the short run, energy demand for both consumption and production is fixed, but in the medium to long run, an increase in the relative price of energy may lead to substitution effects. As a result, macroeconomic aggregates may become less vulnerable to energy shocks. In this regard, wavelet analysis allows us to observe how these relationships change over different time scales. Second, many previous theoretical models that focus on the links between energy prices and inflation have been constructed under the assumption that energy prices are strictly exogenous to macroeconomic variables. This strong hypothesis seems at odds with reality, as growing empirical evidence suggests that not only do energy prices affect the economic system, but expansions in the business cycle also tend to raise energy prices, particularly oil prices (e.g. [18]).

In this chapter, we address these possible shortcomings by combining the estimation of time-domain specifications with the wavelet methodology. Wavelets are a filtering method that decomposes the original series into different frequency components, each accounting for the time series properties of a particular frequency band. Wavelets have several interesting features; however, in our context, we are particularly interested in three main properties. First, wavelets provide a decomposition of the series under consideration into several time series, each associated with different time scales. This makes it possible to observe how the relationship between two time series changes as a function of different time horizons. Second, wavelets can handle nonstationary spells in the time series. This is a particularly important property since macroeconomic time series are notoriously prone to structural breaks (e.g. [4], and the references therein). Third, wavelets are robust to regime shifts and other discontinuities, such as isolated shocks in inflation series. Altogether, these features suggest that wavelets may overcome potential limitations of the time domain modeling framework and provide new insights into the dynamic relationship between inflation and energy prices that the related literature has not yet fully explored.

The empirical analysis is conducted in two steps. In the first stage, we undertake wavelet coherence and phase difference analysis simultaneously to analyze co-movements and causality between inflation, economic uncertainty, and energy prices. We are particularly interested in investigating how these relationships vary across frequencies and over time. In addition, since energy prices also affect consumer prices, we also consider industrial production as a proxy for output growth. This allows us to control for business cycles fluctuations in the euro area countries.

Having analyzed the lead-lag relationships between the time series under consideration, in the second stage of our investigation, we turn our attention to the transmission of shocks that energy price fluctuations impose to the economy system in the euro zone countries. With this target in mind, we build on Alqaralleh et al. [19] and use a procedure that allows to investigate the transmission of shocks between the variables of interest across different time scales. The authors propose to use the maximal overlap discrete wavelet transform to decompose the time series under consideration components associated with different time scale resolutions and use the obtained filtered series to estimate a time varying vector autoregression model (VAR). The resulting procedure not only is robust to structural breaks and nonstationarity, but it allows the variances of the endogenous variables to vary *via* a stochastic volatility Kalman filter.

This work is closely related to previous studies, but it offers a fresh perspective of several long-standing economic issues. First, although based on a different in methodological approach, our chapter builds on the literature that analyses the relationship between energy prices and economic fundamentals. Closely related works are Barsky and Kilian [2]; Hooker [20]; De Gregorio et al. [3] and Kilian [18] where the impact of oil prices on inflation is considered. Other studies have examined the impact of oil prices on consumption (see [21]) and unemployment (see [22]). These works concentrate on the oil prices and generally show a negative effect of increasing oil prices. In this chapter, we extend our interest to non-oil energy price, as well as liquid fuel prices. In the literature, empirical works that include non-oil energy prices are less frequent. However, electricity, gas, heat energy, and solid fuels are employed by households for home heating, and domestic appliances and shocks to these prices historically had a significant impact on the real economy. Also, most of these papers make use of linear models to analyze the impact of energy price shocks on the economic system (e.g. [4, 23–25]). In this chapter, we allow for time-varying parameters and nonlinearity.

Second, we build on the increasing number of works that relate geopolitical risk and economic uncertainty to energy prices shocks. Much of the worlds' energy reserves are located in areas where political turmoil and violence have historically restrained energy production and investment in the energy industry (e.g. Middle east region for oil). Recently, political risks have jeopardized the energy supply from energy rich nations outside regions considered at risk (e.g. Russia and Venezuela) (e.g. [4]). Climate change is also increasingly a source of uncertainty. Hurricanes Rita and Katrina that hit the Gulf Coast of the United States in 2005 provide a good illustration, since they caused enormous disruption to energy production industry in the country (e.g. [26]). It is therefore not surprising that the literature has investigated the relationship between energy price shocks and economic uncertainty (e.g. [7–9, 27]). In this chapter, we build on this literature by considering spillovers between energy prices and the economic policy uncertainty in the time-frequency domain.

Third, the proposed econometric framework builds on a growing literature that uses vector autoregressive models allowing for time-varying volatility when analyzing the effect of energy price shocks (i) on the macroeconomy (e.g. [28]), (ii) on the inflation pass-through, and (iii) on the economic uncertainty (see [27]). While we also control for time-varying volatility and nonlinearity, our main focus is to examine how the dynamic relationships between the variables under consideration change across frequencies.

This chapter is organized as follows. Section 2 describes the data used in this application. Section 3 deals with the wavelet coherence and phase differences empirical investigation along with a brief explanation of the methodology under use. Section 4 describes the main model and presents the empirical results. Finally, Section 5 presents some concluding remarks.

2. Data

The data used in this study consist of monthly data starting from January 1997 to August 2022 for aggregated data on the group of the 19 member states of the European Union on energy prices, inflation rate, industrial production, and economic policy uncertainty.

The inflation is a harmonized consumer price index, (INF_t), and for the 19 member states, the energy prices are a harmonized energy price index, (ENP_t) that includes electricity, gas, liquid fuels, solid fuels, heat energy, and fuels and lubricants for personal transport equipment. To account for energy demand, the harmonized industrial production index, (IP_t), is also considered. Finally, to account for economic uncertainty, we consider the European Economic Policy Uncertainty Index, ($EEPU_t$), suggested by Baker et al. [29].[1]

Figure 1 displays the evolution of the energy price index since 1997, along with the other series under consideration. A visual inspection of **Figure 1** reveals two common structural breaks in the period under consideration, the first occurring in 2009 and the second in 2020. The first break is clearly related to the global financial crisis that occurred in the summer 2007. As it appears from the plot of the industrial production and the inflation series, this period was preceded by a long period of economic expansion. During this period, energy prices sharply increased and became substantially more volatile. This surge was largely driven by oil price due to high demand from China and the Middle East (see [4]).

The second break is related to geopolitical events such as the Covid-19 pandemic and the Russian invasion of Ukraine. As it appears from **Figure 1**, when the pandemic hit the world in 2019, the industrial production collapsed, and the sharp decline of economic activity led to an abrupt decline of energy prices. The severity of the recession pushed inflation and energy prices down. From summer 2020, however, industrial production and energy prices started to recover, mainly due to strong

[1] Note that alternative measures of economic uncertainty are available. However, the EEPU index is a comprehensive measure based on a range of indicators, including the frequency of newspaper references to policy uncertainty. Specifically, the index tracks search results for articles containing terms related to economic policy uncertainty. Since inflation is strongly influenced by economic policy uncertainty, the EEPU index is the most appropriate for our application. Furthermore, Baker et al. [29] evaluated the EEPU index against other news-based measures of uncertainty and found that it outperformed its competitors.

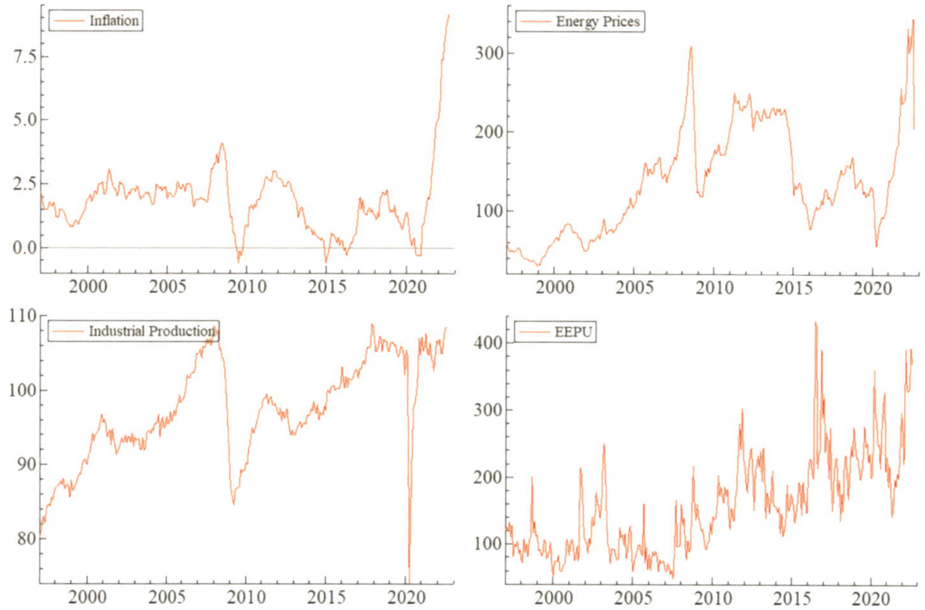

Figure 1.
Monthly data for the harmonized the consumer price index related to the 19 states members of the euro area, harmonized energy price index, the harmonized industrial production index, and the European Economic Policy Uncertainty Index.

demand following the easing of the lock downs. From **Figure 1**, it appears that the industrial production increase was associated with a sharp increase of inflation in the euro zone. Throughout the years 2020 and 2021, the strong economic recovery increased energy demand, and by the summer of 2021, industrial production, energy prices, and inflation levels were back to pre-pandemic levels. As it appears from **Figure 1**, following the Russian invasion of Ukraine in February 2022, energy prices and inflation sharply increased.

From **Figure 1**, it is clear that the EEPU index is able to replicate most of the geopolitical events that increased economic uncertainty during the period under consideration such as the 9/11 terrorist attacks, the U.S.-led invasion of Iraq in 2003, the global financial crisis in 2008–2009, the Brexit referendum in June 2016 in the United Kingdom, the Covid-19 pandemic outbreak in 2019, and the Russian invasion of Ukraine in 2022. Overall, the EEPU index features two different regimes: a downward trend from 1997 to May 2007 and a steady increase from June 2007 onward with the index increasing sharply starting from the Covid-19 outbreak.

3. Energy prices and inflation dynamics

3.1 Wavelet coherence analysis

The purpose of this section is to briefly introduce the concepts and definitions in relation to the wavelet analysis conducted in this study. For a detailed review of the mathematical properties of wavelet, we refer the reader to the discussion in Percival and Walden [30] or Gençay et al. [31].

For the stochastic process y_t the continuous wavelet transforms (CWT) is defined as the following integral

$$W(\tau, s) = \int_{-\infty}^{\infty} y_t \frac{1}{\sqrt{|s|}} \psi^* \left(\frac{t-\tau}{s} \right) dt, \tag{1}$$

where $s, \tau \in R$ are functions of variables that denote the scale and the position, respectively. The asterisk (*) denotes the complex conjugation function of the basic wavelet

$$\psi_{\tau,s}(t) = \frac{1}{\sqrt{|s|}} \psi \left(\frac{t-\tau}{s} \right). \tag{2}$$

In the literature, several wavelet functions have been used for the continuous wavelet transform. In this application, we use the Morlet wavelet defined as

$$\psi(t) = \pi^{-\frac{1}{4}} e^{-i w_{0,t}} e^{\frac{-t^2}{2}}. \tag{3}$$

In Eq. (1), the wavelet $\psi(t)$ must satisfy the admissibility condition

$$\int_{-\infty}^{\infty} \psi(t) dt = 0, \tag{4}$$

and the orthogonality property

$$\int_{-\infty}^{\infty} |\psi(t)|^2 dt = 1. \tag{5}$$

To investigate the pairwise time-frequency relationship between two time series $y_{i,t}$ and $y_{j,t}$, we can investigate the localized covariance by considering the cross-wavelet transform defined as the product of transforms

$$W_{y_{i,t} y_{j,t}}(\tau, s) = W_{y_{i,t}}(\tau, s) W^*_{y_{j,t}}(\tau, s). \tag{6}$$

where the asterisk (*) denotes the complex conjugation function of $W^*_{y_{j,t}}(\tau, s)$. From Eq. (3), the squared wavelet coherence between $y_{i,t}$ and $y_{j,t}$ is given by

$$R^2(\tau, s) = \frac{\left| S\left(\frac{1}{s} W_{y_i y_j}(\tau, s) \right) \right|^2}{S\left(\left(\frac{1}{s} \left| W_{y_i}(\tau, s) \right|^2 \right) S\left(\frac{1}{s} \left| W_{y_j}(\tau, s) \right|^2 \right) \right)}, \tag{7}$$

where S is a smoothing function for both time and frequency.

The expression in Eq. (7), which takes values between 0 and 1, can be interpreted as a localized correlation coefficient in the time-frequency domain. In other words, as it can be seen in Eq. (7), coherence is the ratio of two of the squared cross-wavelet transform to the product of two single power spectrums, analogue to the squared of the coefficient of correlation. Broadly speaking, coherence is a correlation measure that indicates how strongly the two variables are related at business cycle frequencies.

From the cross-wavelet transform, the pairwise lead-lag relationship and the sign of the correlation between time series under consideration can be captured by using the phase difference. The phase for wavelets shows any lag or lead relationship between components and can be expressed as the ratio between the imaginary and the real parts of a cross-wavelet transform and is given by

$$\rho_{y_i,y_j}(\tau,s) = \tan^{-1}\left[\frac{Im\left[\left|S\left(\frac{1}{s}\,W_{y_i y_j}(\tau,s)\right)\right|^2\right]}{Re\left[S\left(\left(\frac{1}{s}\left|W_{y_{i,}}(\tau,s)\right|^2\right)\right)S\left(\left(\frac{1}{s}\left|W_{y_j}(\tau,s)\right|^2\right)\right)\right]}\right]. \quad (8)$$

In Eq. (8) the two series are positively correlated if $-\pi/2 \le \rho_{y_i,y_j} \le \pi/2$, and negative correlated otherwise. Moreover, if the phase difference is $[0, \pi/2]$ and $[-\pi, -\pi/2]$, then the first variable leads the second; however, if $[-\pi/2, 0]$ and $[\pi, \pi/2]$, then the opposite is true, and the second variable is leading.

3.2 Empirical results

The results of the wavelet coherence and the phase difference in Eqs. (7) and (8) are displayed in **Figure 2(a–c)**. In these figures, the degree of which series move together over the business cycle is represented by colors with deep red pointing at the higher level of correlation and deep blue at the lowest. To facilitate the interpretation the frequency component, given on the y-axis, it was converted into time units ranging from 2 months to 64 months. Below we refer to as short-, medium-, and long-run time horizons of less than 8 months, and above 16, respectively. In **Figure 2(a–c)**, the yellow contours indicate statistical significance of the wavelet coherence at 5% significant level. In addition, the lead-lag relationship between inflation and the other variables at hand are capture by the black arrows. In particular, arrows pointing ↗ and ↙ indicate that inflation is leading, whereas arrows pointing to the left, ↘ and ↖, indicate that inflation lagging. Also, the horizontal arrows with left and right directions point at positive and negative correlations, respectively.

Figure 2a presents the estimation results of wavelet coherence and the phase analysis for inflation and energy prices during the period under consideration. Looking at the results from **Figure 2a**, it emerges that statistically significant patches of high correlation are detected mostly during turmoil periods, such as the 9/11 terrorist attacks, the U.S.-led invasion of Iraq in 2003, the global financial crisis, and the sovereign debt and banking crises. Moreover, high levels of coherency are mostly observed in the long-run period, pointing to the fact that inflation is likely to respond to economic fundamentals in the euro area. However, high coherency levels can be noticed across all frequencies after the start of the Covid-19 pandemic and after the Russian invasion of Ukraine in February 2022. Looking at the lead-lag relationships, the causality between inflation in the euro area and energy prices shows some distinctive properties. From 1998 to 2004, energy prices lead inflation, mostly in the time scale 32–64 months, although some causality appears in the short-run period short run (i.e. 8-month frequency band). Similarly, from 2005 to 2012, energy prices led the inflation series, but this time mostly in the 16–32 frequency band. During this period, energy prices substantially increased due to demand pressure from OECD emerging economies, China, and the Middle East in particular. High energy demand was not matched by increase supply from OPEC countries. This supply tightness resulted in oil

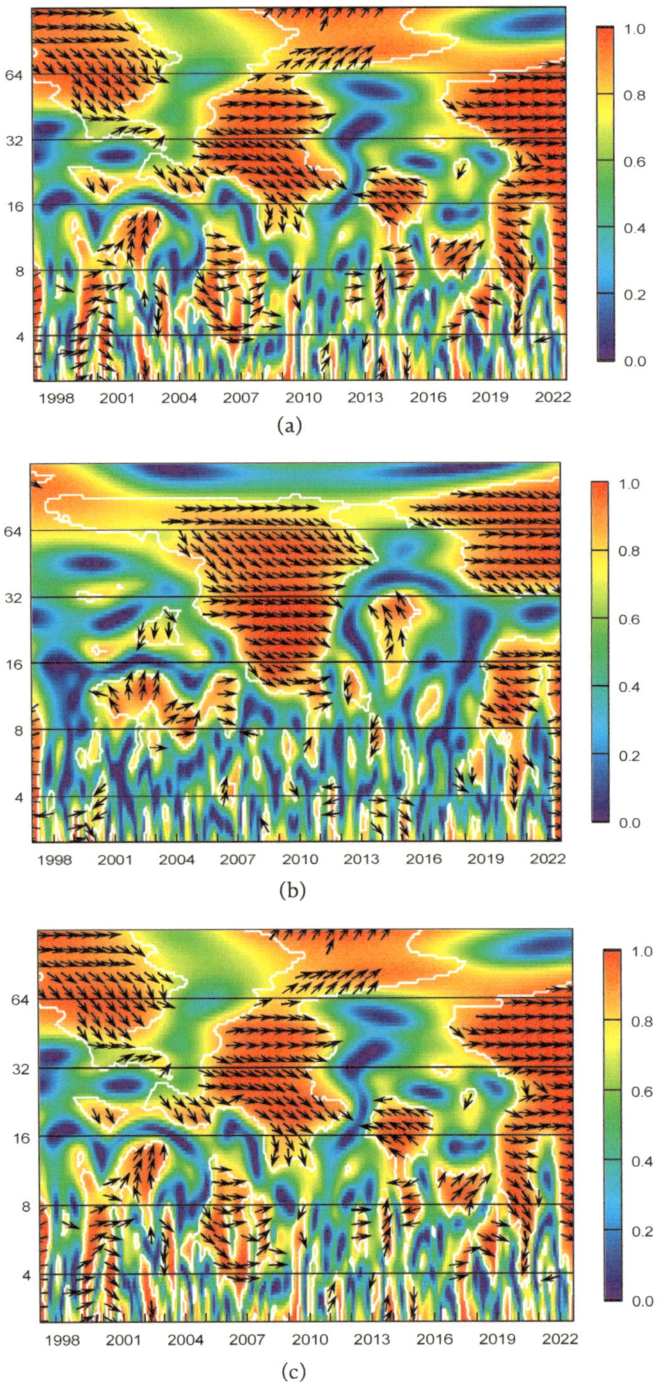

Figure 2.
a. Cross-wavelet transforms between inflation and energy prices in the euro area. b. Cross-wavelet transforms between inflation and industrial production in the euro area. c. Cross-wavelet transforms between inflation and EEPU index in the euro area.

price boom which was suddenly disrupted with the onset of the global financial crisis in 2008 (e.g. [4]). By contrast, the lead-lag relationship inverts, in the short- and medium-run period, from 2013 to 2016, during the sovereign debt crisis when a period of low inflation is observed (see **Figure 1**). In the wake of the global financial crisis, inflation remained low and relatively stable, against the backdrop of the sharpest economic downturn since the Great Depression. As inflation rates kept undershooting 2% target in the euro area countries, the European Central Bank embarked on an unprecedented monetary policy easing. The results in **Figure 2a** seem thus to support the findings in Kilian [18], where it is suggested that there is a reverse causality from macroeconomic variables to energy prices. Finally, from 2019 to 2022, the energy prices significantly led the inflation in the euro area in the short- and medium-run periods. The increase of geopolitical risk and economic uncertainty due to the Covid-19 pandemic and the Russia's invasion of Ukraine in February 2022 are clearly visible in **Figure 2a** (see also **Figure 2c**).

Turning now to the relationship between inflation and industrial production, **Figure 2b** indicates that a high level of coherency between inflation and industrial production exists in the long-run period during times of economic contractions such as the global financial crisis and the sovereign debt crisis. High correlation levels can also be noticed at all frequency scales during the Covid-19 health crisis and since the start of Russia's invasion of Ukraine. Looking at the phase differences, the horizontal arrows indicate positive correlations between the two series with the industrial production leading inflation in the short-, medium-, and long-run periods.

Looking at **Figure 3c**, the red color of the wavelet coherency plot indicates that the Asian financial crisis in 1997 had an impact on inflation over the medium- and long-run periods. Similarly, periods of high correlation are observed during the period 2001–2003, the wake of the 9/11 terrorist attack. Deep red color also appears during the global financial crisis period in 2006 but is particularly evident from the Covid-19 pandemic in 2019 onward.

Coming to the analysis of the phase differences, considering the lead-lag relationships between inflation and the economic uncertainty index, from **Figure 2c**, it appears that the uncertainty index was leading inflation around 2006–2012 in the long-run period. However, the magnitude of the correlation between inflation and the uncertainty index takes an unprecedented level since the start of the Russian invasion of Ukraine, when high levels of correlations are observed across all the time scales under consideration. Geopolitical disorders had a dramatic impact on energy prices since February 2022. In this respect, Bjørnland [32] reports that during the first 2 weeks of the war, Brent prices—the European oil benchmark—increased by more than 25%. Likewise, at the end of March 2022, European gas prices were around 580% higher than a year earlier. The impact of this huge energy prices increases along with the high level of political uncertainty can clearly be seen in **Figure 2c**.

4. The transmission channel of shocks

The wavelet coherence analysis presented in the previous section illustrates how inflation, energy prices, industrial production, and economic uncertainty interact overtime and across frequencies. The wavelet coherency analysis revealed that there is

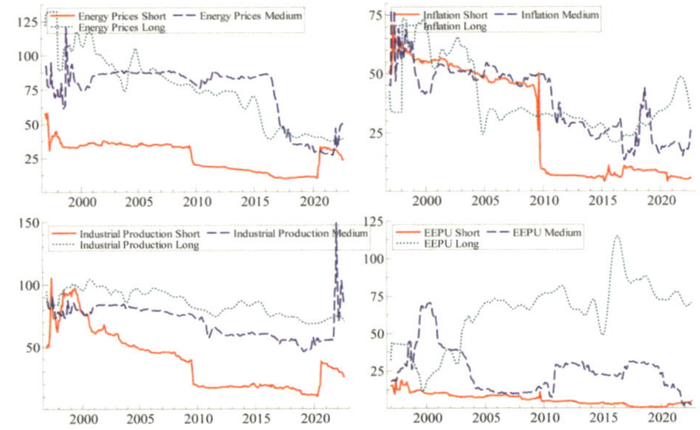

(a)

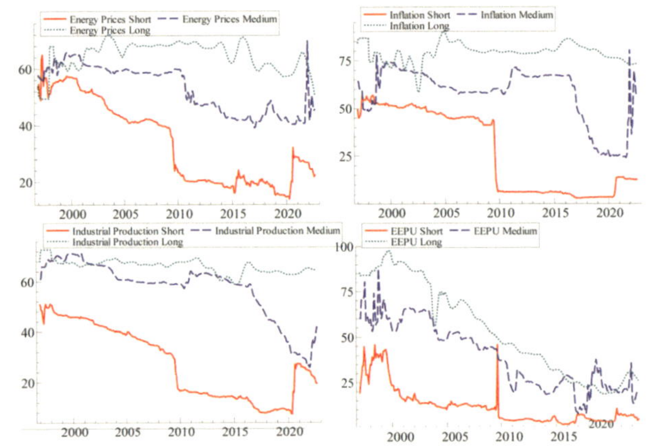

(b)

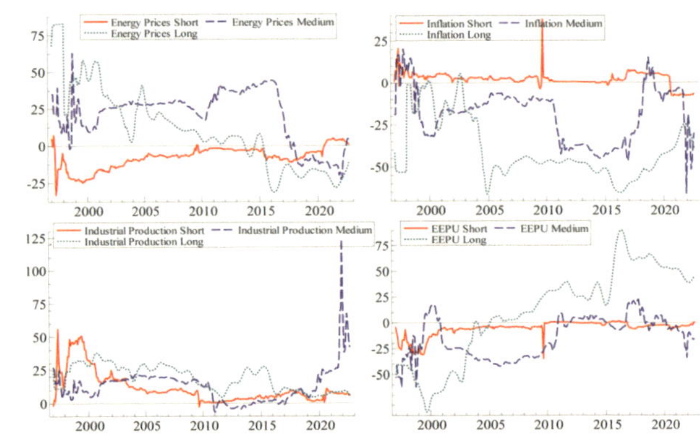

(c)

Figure 3.
a. Total shock transmission To others in the short-, medium-, and long-run periods. b. Total shock transmission FROM others in the short-, medium-, and long-run periods. c. Net total shock transmission in the short-, medium-, and long-run periods.

strong correlation between inflation and the other variables under consideration, in particular during turmoil periods. However, the strength of the relationship not only changes across the business cycle, but also across frequencies. In general, high and statistically significant correlations are found in the medium- and long-run periods, namely for fluctuations that last for more than 16 months. Concerning the shorter-run movements, the wavelet coherence analysis highlights episodes where the links have been temporarily stronger. From the analysis of lead-lag relationship, we can conclude that energy prices lead inflation and industrial production at long-term movements; however, for other time-frequency regions, the picture is less defined, with phase differences oscillating between lag and lead, although the lead relationship seems to be prevalent.

In this section, we turn our attention to the dynamic relations between the variables under investigation with a particular interest on the impact of unexpected energy prices shocks on inflation. With this target in mind, we consider the wavelet time-varying parameter vector autoregression (WTVP-VAR) procedure proposed in the study by Alquarelleh et al. [19]. The authors suggest using the maximal overlap discrete wavelet transform (MODWT) combined to a Bayesian time-varying VAR model to investigate the propagation of shocks within the variables of interest over different time scales. The implementation of the estimation procedure involves undertaking two steps that are briefly described below.

4.1 First step: Discrete wavelet decomposition

In the previous section, the continuous wavelet transform was used to decompose the stochastic processes under consideration. The CWT was obtained by convolving a signal with an infinite number of functions, generated by translating (t) and scaling the Morlet scaling function. The resulting two-dimensional transform has parameters that vary continuously. The advantage of CWT is that it delivers smoothly varying local frequency and scales, and therefore, important signal features of the series are easily detected. However, for the purpose of estimating a VAR model, a discrete wavelet transform (DWT) needs to be used, as in this case, the decomposition provides a discrete representation of the signal in both the time and frequency domains.[2]

The MODWT can be used to decompose the series under investigation into different time scales as

$$y(t) = \sum_k s_{m,k}\phi_{m,k}(t) + \sum_k d_{m,k}\psi_{m,k}(t) + \sum_k d_{i-1,k}\psi_{m-1,k}(t) + \dots + \sum_k d_{1,k}\psi_{m-1,k}(t), \tag{9}$$

where $\phi_{m,k}$ and $\psi_{m,k}$ are two basic functions that capture the low-frequency and high-frequency oscillations of the series, respectively. In Eq. (9), m represents the number of time scales, and k ranges from 1 to the number of coefficients in the specified components. The coefficients $s_{m,k}$ and $d_{m,k}$ are the wavelet transform coefficients given by

$$s_{m,k} = \int f(t)\phi_{m,k}(t)dt, \tag{10}$$

[2] MODWT has often been used in the related literature since it has some desirable properties. For example, the MODWT wavelet transform is shift invariant in the sense that circularly shifting the time series by any amount will also shift the MODWT wavelet scaling and details coefficients by a corresponding amount. Also, the MODWT details and smooth coefficients are associated with a zero phase filters and thus allow to replicate feature of the original series in the multiresolution analysis. For more details see Percival and Walden [30].

$$d_{m,k} = \int f(t)\psi_{m,k}(t)dt \text{ for } m = 1, 2, \ldots M. \quad (11)$$

defined as

$$\phi_{m,k} = 2^{-m/2}\phi\left(\frac{t - 2^m k}{2^m}\right), \quad (12)$$

$\psi_{m,k} = 2^{-m/2}\psi\left(\frac{t-2^m k}{2^m}\right)$.

In Eq. (9), the transform coefficient $s_{m,k}$ represents the smooth coefficient that captures the trend, whereas the detail coefficients $s_{m,k}, \ldots, s_{1,k}$ capture the higher frequency oscillations. Therefore, the wavelet approximation of the original series can be expressed as the smooth and the details signals

$$y_t = S_M + D_M + D_{M-1} + \ldots + D_1.m = 1, \ldots, M \quad (13)$$

where the frequency components D_M are listed in the order of increasingly finer components, and S_M is a smooth coefficient that captures the deviation from the trend (for details on the MODWT, see also [30]).

4.2 Second step: WTVP-VAR estimation

The multiresolution decomposition obtained in the first step allows us to analyze impact of shocks at the short-, medium-, and long-term impact by shocks arising at different time scales. With this target in mind, the filtered series obtained in the first step are used to estimate the TVP-VAR model in the time-frequency framework. The model can be written as follows

$$z_t = \beta_t z_{t-1} + \epsilon_t; \epsilon_t | F_{t-1} \sim N(0, S_t), \quad (14)$$

$$vec(\beta_t) = vec(\beta_{t-1}) + v_t, v_t | F_{t-1} \sim N(0, R_t), \quad (15)$$

where z_t and $z_{t-1} = \left[y_{t-1}, \ldots, y_{t-p}\right]'$ represent $N \times 1$ and $P \times 1$ dimensional vectors, respectively. β_t is an $N \times$ Np dimensional time-varying coefficient matrix, ϵ_t is an $N \times 1$ dimensional error disturbance vector with an $N \times N$ time-varying variance-covariance matrix S_t, $vec(\beta_t)$ and v_t are N^2p $\times$ 1 dimensional vectors, and R_t is an N^2p $\times N^2 p$ dimensional matrix.

For each frequency band, the model in Eqs. (14) and (15) can be estimated by Kalman filter allowing for time-varying variance covariance matrices, and the estimated parameters can be used to calculate the generalized forecast error variance decompositions (GFEVD) (see [27, 33, 34]). In particular, from the estimated parameters in Eqs. (14) and (15), the general impulse response function (GIRF) can be obtained by transforming the TVP-VAR using the Wold representation theorem to obtain a TVP-VMA:

$$z_t = \sum_{j=0}^{\infty} L' W_t^j L \epsilon_{t-j}, \quad (16)$$

$$z_t = \sum_{j=0}^{\infty} A_{it} \epsilon_{t-j}, \quad (17)$$

where $L = [I_N, \dots, 0_p]'$ is an $Np \times N$ dimensional matrix, $W = [\beta_t; I_{N(p-1)}, 0_{N(p-1)\times N}]$ is an $Np \times Np$ dimensional matrix, and A_{it} is an $N \times N$ dimensional matrix.

From Eq. (16), the generalized impulse response functions (GIRFs) and generalized forecast error variance decompositions (GFEVDs) can be calculated. For each variable i, the GIRF represents the reactions of all variables following a shock in i and computed as follows

$$GIRF_t\ (K, \delta_{j,t}, F_{t-1}) = E(y_{t+K}|\epsilon_{j,t} = \delta_{j,t}, F_{t-1}) - E(Y_{t+K}|\ F_{t-1}), \tag{18}$$

$$\psi_{j,t}(K) = \frac{A_{K,t}S_t\ \epsilon_{j,t}}{\sqrt{S_{jj,t}}} \frac{\delta_{j,t}}{\sqrt{S_{jj,t}}}, \delta_{j,t} = \sqrt{S_{jj,t}}, \tag{19}$$

$$\psi_{j,t}(K) = \frac{A_{K,t}S_t\ \epsilon_{j,t}}{\sqrt{S_{jj,t}}}, \tag{20}$$

where $\psi_{j,t}$ represents the GIRFs of variable j, K represents the forecast horizon, $\delta_{j,t}$ the selection vector with one on the jth position and zero otherwise, and F_{t-1} the information set until $t-1$. Similarly, from Eq. (16), GFEVD can be computed as follows:

$$\tilde{\Phi}_{ij,t}(K) = \frac{\sum_{t=1}^{K-1} \psi_{j,t}^{2,g}}{\sum_{j=1}^{N} \sum_{t=1}^{K-1} \psi_{j,t}^{2,g}} \cdot \sum_{j=1}^{N} \tilde{\Phi}_{ij,t}(K) = 1\ and \sum_{j=1}^{N} N_{ij,t}(K) = N \tag{21}$$

Eq. (21) can be used to estimate the influence that the j variable has on variable i in terms of its forecast error variance share. By normalizing the variance shares, one can explore how a shock to the i variable spills over to all the other variables of the system through the total shock transmission index, which can be constructed as

$$C_t(K) = \frac{\sum_{i,j=1, i\neq j}^{N} \tilde{\Phi}_{ij,t}(K)}{N} * 100. \tag{22}$$

More interesting is to analyze the direction of the shock transmission due to volatility spillover. The method under consideration considers three aspects of this direction.

First, total shock transmission *TO others,* given as

$$C_{i\to j,t}(K) = \frac{\sum_{i,j=1, i\neq j}^{N} \tilde{\Phi}_{ij,t}(K)}{\sum_{j=1}^{N} \tilde{\Phi}^{g}{}_{ji,t}\ (K)} * 100. \tag{23}$$

Eq. (23) gives information on how a shock to a variable i is transmitted to all the other j variables. Similarly, we can calculate the total shock received *FROM others,* which gives information on how shocks to the j variables are transmitted to the i variable. This is given as

$$C_{i\leftarrow j,t}(K) = \frac{\sum_{i,j=1, i\neq j}^{N} \tilde{\Phi}_{ij,t}(K)}{\sum_{j=1}^{N} \tilde{\Phi}_{ij,t}(K)} * 100. \tag{24}$$

Last, by subtracting Eq. (23) from Eq. (22) to obtain the *net directional shock transmission* as follows:

$$C_{i,t}(K) = C_{i\to j,t}(K) - C_{i\leftarrow j,t}(K). \tag{25}$$

Eq. (25) illustrates the *NET* magnitude of the impact that i variable has on the analyzed network. Thus, a positive value of Eq. (25) implies that the shock that the i variable causes to the other variables within the network is greater than the one that the network produces on i (i.e. the variable i is a net shock producer in the economic system under consideration), whereas a negative value implies that shocks induced to i by the network under consideration are greater than the ones caused by the i variable to the network (i.e. the variable i is a net shock receiver).

4.3 Empirical results

4.3.1 Static analysis of multi-scale shock transmission

In **Table 1**, panels (a), (b), and (c), we report the estimation results for the spillover effects defined in Eqs. (23) and (24) for the short-, medium-, and long-run periods, respectively. The estimation results were obtained estimating a VAR(1) model with lag order chosen on the base of the Schartz information criterion.

In **Table 1**, for each panel, the diagonal elements represent the intra-variable spillovers obtained from the generalized forecast error variance decomposition matrix. Conversely, the off-diagonal elements of the table report the 12-month step ahead error variance that is explained by shocks to the other variables in the network. Therefore, it is a measure of the extent to which each variable in the network receives or transmits shocks within the network. In addition, in panels (a), (b), and (c), the row sums labeled as "FROM" (see Eq. (24)) report the total spillovers received by the variable in the respective row. Similarly, the column sum, labeled as "TO" (see Eq. (23)), reports the total spillovers transmitted by the variable in the respective column. In the last row of each panel, the *NET* spillover is calculated as the difference between each variable's (off-diagonal) column sum and the same variable's row sum (see Eq. (25)).

Looking at the results in **Table 1(a)–(c)**, we observe that no matter the time scale under consideration, the own-series spillovers explain the highest shares of forecast error variance, as the diagonal elements receive higher values compared to the off-diagonal elements. The shares of intra-variable spillovers give important information on shocks to the second moment of each variable of the WTVP-VAR and can be interpreted as intra-variable uncertainty shocks. In this respect, comparing the diagonal elements of the estimation results presented in panels a), b), and c) gives us information on the persistence of these intra-variable uncertainty shocks. For example, from panel a) we can infer that shocks to inflation explain 72.99% of the average 12-month ahead forecast error variance in the short run, and therefore, inflation shocks in the short run depend more on inflation uncertainty than the transmission of shocks from the other variables in the system. However, this picture changes in the medium- and long-run periods when the estimated impact of the intra-variable uncertainty shock reduces to 42.5 and 20.7%, respectively (see panels (b) and (c)). These results are in line with the seminal work by Cukierman and Meltzer [35] where uncertainty about future inflation also affects the average rate of inflation (see, inter alia, [1, 7, 8, 21]).

	EEUI	IP	ENP	INF	FROM
Panel A: Short term (< 8 months)					
EEUI	88.35	5.02	3.58	3.06	11.65
IP	2.32	72.26	13.25	12.18	27.74
ENP	2.01	17.92	66.08	13.99	33.92
INF	1.61	16.07	9.33	72.99	27.01
TO	5.94	39	26.15	29.23	100.33
NET	−5.71	11.26	−7.77	2.21	
Panel B: Medium term (8–16 months)					
EEUI	60.58	12.02	15.81	11.59	39.42
IP	9.55	42.55	33.77	14.14	57.45
ENP	10.45	29.31	47.17	13.08	52.83
INF	4.64	29.15	23.72	42.5	57.5
TO	24.63	70.47	73.3	38.8	207.21
NET	−14.79	13.01	20.47	−18.7	
Panel C: Long term (over 16 months)					
EEUI	46.47	20.34	23.22	9.97	53.53
IP	25.95	33.92	25.49	14.64	66.08
ENP	18.35	31.67	35.36	14.62	64.64
INF	19.62	33.13	26.55	20.7	79.3
TO	63.91	85.14	75.26	39.23	263.54
NET	10.38	19.06	10.62	−40.07	

Note: The estimation results are based on a WTVP-VAR model with a lag length of order one and a 12-month-step-ahead generalized forecast error variance decomposition. Short, medium, and long term is referred to as the obtained series from the m-level multi-resolution decomposition.

Table 1.
Shocks transmission by time scale.

Looking at the shares of intra-variable spillover of energy prices, from the fourth column of **Table 1**, it appears that shocks to energy prices explain 66.08% of the average 12-month ahead forecast error variance in the short-run period, and 47.17, 35.36% in the medium and long run, respectively, signaling that uncertainty shocks to energy prices in the euro area are rather persistent. These findings may have important implications for policymakers, since according to a well-established literature, the negative contribution of energy price shocks to the economic activity is greater when the persistence of energy price shocks is high, as it becomes more difficult for central banks to stabilize the economy (see [36]). Similarly, from **Table 1**, it appears that uncertainty shocks are rather persistent, since a shock that explains 88.35% of the average 12-month ahead forecast error variance within the 8-month period still explains 46.7% of the variability in the (up to) 32-month period.

Turning our attention at the off-diagonal elements of **Table 1**, looking at the fourth column, we see that energy price shocks explain 9.33% of the 12-month-ahead forecast error variance on inflation in the short run. However, in the medium- and long-run periods, these shares increase to 23.72 and 26.55, respectively. As expected, energy price shocks also are transmitted to industrial production. However, the impact is mostly felt in the medium-run period, when the spillover share is 33.77%. These results imply that the speed of propagation of energy price shocks to the real economy is relatively slow. In the short term, the ability of firms to react to energy price shocks by substituting energy with other inputs is rather limited, and this could potentially explain the estimated results in **Table 1**.

Coming to the spillover effects of energy price shocks to economic uncertainty, from columns two and four, it appears that there is bidirectional relationship between and economic uncertainty and energy price shocks. On one hand, energy price shocks increase economic uncertainty, especially in relation to monetary policy response. On the other hand, there is uncertainty in relation to the economic outlook negatively investment and output, which in turn causes a reduction of output levels and energy demand. Looking at the speed of propagation, from **Table 1**, it appears that the transmission of uncertainty shocks reaches its maximum only after 16 months.

Turning to the impact of economic uncertainty, from column two, it appears that there is an important spillover effect from economic uncertainty to the energy prices, inflation, and industrial production. For example, a shock to the uncertainty index explains 2.32% of the average 12-month-ahead forecast error variance of the industrial production in the short run, 9.55% in the medium run period, and 25.95% in the long run. Similarly, it explains 18.35 and 19.62% of the average shares of energy prices and inflation in the long-run period.

Overall, the average "net" spillovers demonstrate that energy prices are net shock transmitters in the medium and long run, but shock receivers in the short run. These results support the view that the speed of the shock transmission to the economic activity is relatively slow, as it takes more than 8 months for the pass-through effect to the other variables of the system. On the opposite, the estimated "net" spillover results for inflation highlight that inflation is a shock transmitter in the short run, but shock receiver in the medium and long run. From column two, we can infer that shocks to industrial production are also an important source of instability, since from the estimated "net" effect, (see the third column in **Table 1**), it appears that the difference between the directional spillover transmitted to other and the directional spillover received from others is always positive, no matter the time scale into consideration. Finally, looking at the "net" estimation results for the EEPU index, it is clear that the economic system responds to geopolitical events and uncertainty in the economic system, but the diffusion of a shock is rather slow, since the EEPU index becomes a net receiver only after 16 months.

4.4 Dynamic analysis of multi-scale shock transmission

The average results in **Table 1** provide important information on the transmission channels of energy shocks and the propagation mechanism of these shocks throughout the economic system. We now turn our attention to the time-varying interpretation of the spillover plot based on the time-varying estimates of the various connectedness measures. Here, we are particularly interested in investigating how geopolitical shocks affect the variables under investigation.

Figure 3(a–c) present the results provided by the WTVP-VAR for the each of the time series under consideration. In particular, **Figure 3a** provides, for each time period, the estimated dynamic TO total directional spillover, where we look at the case where variable *i* transmits its shock to all other variables. **Figure 3b** plots, for each time scales, the estimated results for the FROM dynamic total directional spillover, where we look at the shocks that variable *i* receives it from the other variables in the system. Finally, the time-varying results for the NET spillovers are presented in **Figure 3c**. As before, a positive (negative) value in **Figure 3c** illustrates that variable *i* is driving the network more (less) than being driven by it. Hence, variable *i* is a net transmitter (receiver). Note that in the interest of brevity, the time-varying estimation result for the "Total" has been omitted, but the results are available upon request.

According to the results plotted, we observe that the estimated time-varying spillover indexes are very responsive to downturn the economy activity and geopolitical events. In general, the results in **Figure 3** support the results from the wavelet coherence analysis presented in **Figure 2** and **Table 1**.

5. Conclusion

In this chapter, we assess the relationship between energy prices, economic uncertainty, industrial production, and inflation in the euro area region. The main findings of this application can be summarized as follows. First, significant correlation is found between energy prices and inflation. Statistically significant patches of high correlation are detected mostly during turmoil periods such as the 9/11 terrorist attacks, the U.S.-led invasion of Iraq in 2003, the global financial crisis, and the sovereign debt and banking crises. In particular, high coherency levels can be noticed across all frequencies after the start of the Covid-19 pandemic and the Russian invasion of Ukraine in February 2022. Looking at the lead-lag relationships, it is found that energy prices lead inflation in most cases. However, patches of reverse causality are also detected, especially in the medium-run period. These results support the strand of literature claiming that energy prices are endogenous with respect to macroeconomic conditions (e.g. [2, 36, 37]). Endogeneity in our case refers to the fact that not only do energy prices affect the economy, but there is also reverse causality from macroeconomic aggregates to the price of energy.

Second, looking at the estimated shares of intra-variable spillovers of energy prices, it appears that the impact of uncertainty shocks to energy markets is rather persistent. These results have important implications for policymakers, since the impact of energy price changes depends not only on their driving forces, but also on their persistence and how they are absorbed by the economy. On the contrary, inflation uncertainty is much less persistent, since shocks on inflation explain large part of the average 12-month ahead forecast error variance in the short-run period, but much less in the medium- and long-run periods.

Third, energy prices are important shock transmitters in the network of variables under consideration. However, the magnitude of the transmission changes over time and across frequencies. In particular, our estimation results suggest that energy price shocks are transmitted to industrial production and inflation. However, the impact is mostly felt after 8 months from the shocks in both cases. These results imply that there is a pass-through effect, but the speed of propagation of energy price shocks to the economic system in the euro zone is rather slow.

Finally, the estimation results suggest that geopolitical and economic uncertainty are major shock transmitters. It appears that there is bidirectional relationship between and economic uncertainty and energy price shocks. On one hand, energy price shocks increase economic uncertainty, especially in relation to monetary policy response. On the other hand, uncertainty surrounding economic conditions negatively affects firms' investment and output decisions, which in turn causes a reduction of output levels and a downward pressure on energy (see [38]). Also, the estimation results from the WTVP-VAR reveal that the diffusion of a shock to the economic system is rather slow, since the EEPU index becomes a net shock transmitter only after 16 months from the shocks.

Acknowledgements

Alessandra Canepa acknowledges financial support from the Italian PRIN 2022 grant "Methodological and computational issues in large-scale time series models for economics and finance" (20223725WE).

JEL classification

C01; C4; C22; E6; Q4

Author details

Huthaifa Sameeh Alqaralleh[1] and Alessandra Canepa[2,3]*

1 Department of Economics, Business and Finance, Mutah University, Karak, Jordan

2 Department of Economic and Statistics Cognetti De Martiis, University of Turin, Turin, Italy

3 Department of Economics and Finance, Brunel University London, United Kingdom

*Address all correspondence to: alessandra.canepa@unito.it

References

[1] Hamilton JD. Oil and the macroeconomy since world war II. Journal of Political Economy. 1983;**91**: 228-248

[2] Barsky RB, Kilian L. Oil and the macroeconomy since the 1970s. Journal of Economic Perspectives. 2004; **18**:115-134

[3] De Gregorio J, Landerretche O, Neilson C. Another pass-through bites the dust? Oil prices and inflation. Economia. 2007;**2**:155-208

[4] Hamilton JD. Causes and Consequences of the Oil Shock of 2007–08. Vol. Spring. Brookings Papers on Economic Activity; 2009. pp. 215-261

[5] Kilian L, Park C. The impact of oil price shocks on the U.S. stock market. International Economic Review. 2009; **50**:1267-1287

[6] Kilian L, Lewis LT. Does the fed respond to oil price shocks? Economic Journal. 2011;**121**:1047-1072

[7] Montoro C. Oil shocks and optimal monetary policy. Macroeconomic Dynamics. 2012;**16**:240-277

[8] Natal J-M. Monetary policy responses to oil price shocks. Journal of Money, Credit, and Banking. 2012;**44**:53-101

[9] Pieschacón A. The value of fiscal discipline for oil-exporting countries. Journal of Monetary Economics. 2012;**59**: 250-268

[10] Segal P. Oil price shocks and the macroeconomy. Oxford Review of Economic Policy. 2011;**27**:169-185

[11] Álvarez LJ, Hurtado S, Sánchez I, Thomas T. The impact of oil price changes on Spanish and euro area consumer price inflation. Economic Modelling. 2011;**28**:422-431

[12] Blanchard O, Galí J. The macroeconomic effects of oil price shocks: Why are the 2000s so different from the 1970s? In: International Dimensions of Monetary Policy. Cambridge, MA: NBER, University of Chicago Press; 2010

[13] Chen SS. Oil price pass-through into inflation. Energy Economics. 2009;**31**: 126-133

[14] Feyrer J, Mansur ET, Sacerdote B. Geographic dispersion of economic shocks: Evidence from the fracking revolution. American Economic Review. 2017;**107**:1313-1334

[15] Gelos G, Ustyugova Y. Inflation responses to commodity price shocks– how and why do countries differ. Journal of International Money and Finance. 2017;**72**:28-47

[16] Gilje E, Ready R, Roussanov N. Fracking, Drilling, and Asset Pricing: Estimating the Economic Benefits of the Shale Revolution. NBER Working Paper no. 22914. Cambridge, MA: National Bureau of Economic Research; 2016

[17] Habermeier K, Kokenyne A, Veyrune R, Anderson H. Inflation Pressures and Monetary Policy Options in Emerging and Developing Countries-A Cross Regional Perspective. Washington, D.C: IMF Working Papers; 2009

[18] Kilian L. The economic effects of energy price shocks. Journal of Economic Literature. 2008;**46**:871-909

[19] Alqaralleh H, Canepa A, Salah Uddin G. Dynamic relations between housing markets, stock markets, and uncertainty in global cities: A time-frequency approach.

The North American Journal of Economics and Finance. 2023;**68**(C):101950

[20] Hooker MA. Are oil shocks inflationary? Asymmetric and nonlinear specifications versus changes in regime. Journal of Money, Credit and Banking. 2002;**34**:540-561

[21] Ferderer JP. Oil price volatility and the macroeconomy. Journal of Macroeconomics. 1996;**18**:1-26

[22] Caruth AA, Hooker MA, Oswald AJ. Unemployment equilibria and input prices: Theory and evidence from the United States. Review of Economics and Statistics. 1998;**80**:621-628

[23] Aastveit KA. Oil price shocks in a data-rich environment. Energy Economics. 2014;**45**:268-279

[24] Edelstein P, Kilian L. How sensitive are consumer expenditures to retail energy prices? Journal of Monetary Economics. 2009;**56**:766-779

[25] Lippi F, Nobili A. Oil and the macroeconomy: A quantitative structural analysis. Journal of the European Economic Association. 2012; **10**:1059-1083

[26] Azhgaliyeva D, Kapsalyamova Z, Mishra R. Oil price shocks and green bonds: An empirical evidence. Energy Economics. 2022;**112**:106108

[27] Antonakakis N, Chatziantoniou I, Filis G. Dynamic spillovers of oil price shocks and economic policy uncertainty. Energy Economics. 2014; **44**:433-447

[28] Baumeister C, Peersman G. Time-varying effects of oil supply shocks on the US economy. American Economic Journal: Macroeconomics. 2013;**5**:1-28

[29] Baker SR, Bloom N, Canes-Wrone B, Davis SJ, Rodden J. Why has US policy uncertainty risen since 1960? American Economic Review. 2014;**104**:56-60

[30] Percival DB, Walden AT. Wavelet Methods for Time Series Analysis (Vol. 4). Cambridge: Cambridge University Press; 2000

[31] Gençay R, Selçuk F, Whitcher BJ. An Introduction to Wavelets and Other Filtering Methods in Finance and Economics. Elsevier; 2001

[32] Bjørnland HC. The effect of rising energy prices amid geopolitical developments and supply disruptions. In: Paper Presented ECB Forum on Central Banking 2022. Frankfort. 2022

[33] Diebold FX, Yilmaz K. Better to give than to receive: Predictive directional measurement of volatility spillovers. International Journal of Forecasting. 2012;**28**:57-66

[34] Koop G, Pesaran M, Potter S. Impulse response analysis in nonlinear multivariate models. Journal of Econometrics. 1996;**74**:119-147

[35] Cukierman A, Meltzer A. A theory of ambiguity, credibility, and inflation under discretion and asymmetric information. Econometrica. 1986;**54**:1099-1128

[36] Bernanke BS, Boivin J, Eliasz P. Measuring the effects of monetary policy: A factor-augmented vector autoregressive (FAVAR) approach. The Quarterly Journal of Economics. 2005; **120**:387-422

[37] Bjørnland HC, Thorsrud LA, Torvik R. Dutch disease dynamics re-considered. European Economic Review. 2019;**119**:411-433

[38] Bloom N. The impact of uncertainty shocks. Econometrica. 2009;**77**:623-685

Chapter 4

Schlömilch Series Based on Integral Transforms

Slobodan B. Tričković and Miomir S. Stanković

Abstract

Schlömilch series was defined originally as a Fourier-type series expansion of twice continuously differentiable function $f(x)$ in the interval $(0, \pi)$ in terms of the Bessel function of the first kind, named after the German mathematician Oskar Schlömilch. Instead of Bessel functions, some functions related to them may appear in the expansion. Also, it is interesting to consider the series with an integral of the Bessel, Struve, or related functions. In place of Bessel or associated functions, we can take a product of these functions multiplied by a trigonometric function. We express the summation formulas as series over the functions Riemann's zeta or Dirichlet eta, lambda, and beta. We can bring these series in closed form for some parameter α values in the denominator.

Keywords: gamma function, Riemann and Hurwitz zeta function, Dirichlet eta, lambda and beta function, harmonic numbers, hypergeometric function, Bessel and Struve function, Fourier, Mellin and Hankel transform, Schlömilch series

1. Introduction

The Erdélyi-Kober operator of fractional order is defined by (see [1, p. 433])

$$P^{(\alpha)}(\varphi; z) = \int_0^1 (1-t^2)^{\alpha-1/2}\varphi(zt)\mathrm{d}t, \quad \mathrm{Re}\ \alpha > -\frac{1}{2}, \tag{1}$$

and for $\alpha = 0$ and $\varphi = \cos$, it yields Poisson's integral representation of the Bessel function $J_0(z)$ (see [2, p. 224, entry 10.9.4 for $\nu = 0$])

$$J_0(z) = \int_0^1 (1-t^2)^{-1/2}\cos(zt)\mathrm{d}t. \tag{2}$$

Rusev used (1) to give a necessary and sufficient condition (see [1, Theorem 1.2]) for a holomorphic function $f(x)$ (a complex-valued function that is differentiable in the vicinity of each point in a domain) to be represented by the series

$$f(x) = \sum_{n=1}^{\infty} a_n J_\nu(nx), \tag{3}$$

where $\nu = 0$. Rayleigh (see [3]) showed that the series Eq. (3) plays a significant role in physics, and they are applicable when one investigates periodic transverse vibrations uniformly distributed in direction through the two dimensions of the membrane.

Except for Bessel functions $J_\nu(nx)$, we can consider Struve $\mathbf{H}_\nu(nx)$ and related functions. Also, we can leverage their integrals designated with $D_\nu(nx)$, which present integral transforms of these functions (see [4–6]). Thus, we deal with the series

$$\sum_{n=1}^{\infty} a_n D_\nu(nx). \tag{4}$$

In addition, we can take a product of these functions multiplied by the sine or cosine function, that is,

$$\sum_{n=1}^{\infty} a_n g(ny) D_\nu(nx) \tag{5}$$

with g presenting one of these trigonometric functions. We call all these series the Schlömilch-type series, and getting at summation formulas for them reduces to finding its trigonometric series representation. So, one regards trigonometric series in a broader sense as a fundamental goal of the investigation.

Apart from following its internal logic and laws based on axioms and abstract notions, mathematics draws on appearances in the real world. To solve problems, it develops its models and returns these solutions to the real world. In the process, it achieves results that someone might use later but is not predictable when. The same can apply to the material offered and presented here.

In paper [7], authors used a simple quantum mechanical model for obtaining new infinite sums, including products of Bessel functions, applying our results from [8] (Schlömilch-like infinite series [9]), as well as those containing generalized hypergeometric functions, which, to the best of our knowledge, have not been calculated yet. One of the simplest solvable quantum mechanical models, also used in [10], describes a particle of mass m moving in the infinite potential well.

1.1 Integral transforms

We recollect definitions and some basic properties of the integral transforms, and we shall apply in the sequel.

An integral transform maps a function into another function space via integration, where we manipulate some properties easily. The inverse transform generally maps back the image to the original domain.

There are classes of problems that are not easy to solve in their original representations. An integral transform maps an equation from its domain into another one. Solving the equation in the target domain can be much easier than handling a solution in the original. Then, it maps back with the inverse of the integral transform.

A general integral transform $\mathcal{T}$ of a complex-valued function $f(x)$ of a real variable x is defined by

$$\mathcal{T}\{\phi(t); s\} = \int_a^b \phi(t) K(s,t)\, \mathrm{d}t, \tag{6}$$

providing that the above integral exists, s is a complex variable, $K(s,t)$ a complex-valued function, with a and b as are real numbers, where can be $a = -\infty, b = +\infty$. The most used transforms of the form (6) are as follows:

1° Fourier, if $K(s,t) = \frac{1}{\sqrt{2\pi}} e^{ist}$, $a = -\infty, b = +\infty$,

2° Mellin, if $K(s,t) = t^{s-1}$, $a = 0, b = +\infty$,

3° Hankel, if $K(s,t) = tJ_\nu(st)$, $a = -\infty, b = +\infty$.

One of the basic and useful properties of the integral transform $\mathcal{T}$ (is)

$$\mathcal{T}\{c_1\phi_1(t) + c_2\phi_2(t); s\} = c_1\mathcal{T}\{\phi_1(t); s\} + c_2\mathcal{T}\{\phi_2(t); s\}. \tag{7}$$

for arbitrary constants c_1, c_2. The property (7) is called linearity.

1.1.1 The Fourier transform

The Fourier transform is defined (see [11])

$$F(s) = \mathcal{F}\{\phi(x); s\} = \int_{-\infty}^{\infty} \phi(x) e^{ixs}\, \mathrm{d}x. \tag{8}$$

Suppose that ϕ and ϕ' are piecewise continuous on $(-\infty, \infty)$, meaning that both are continuous in any finite interval except possibly for a finite number of jump discontinuities. Suppose further that $|f|$ is integrable. Then, at each point x where $\phi(x)$ is continuous, the inverse Fourier transform one obtains is as follows (see [12])

$$\phi(x) = \mathcal{F}^{-1}\{F(s); x\} = \frac{1}{2\pi} \int_{-\infty}^{\infty} F(s) e^{-ixs}\, \mathrm{d}s.$$

The Fourier sine and cosine transforms denoted by $\mathcal{T}$ (see [13]) are forms of the Fourier transform that do not use complex numbers or require negative frequency.

$$T(x) = \left\{ \begin{array}{c} S(x) \\ C(x) \end{array} \right\} = \mathcal{T}\{\phi(y); x\} = \int_0^{\infty} \phi(y) \left\{ \begin{array}{c} \sin xy \\ \cos xy \end{array} \right\} \mathrm{d}y, \tag{9}$$

originally used by Joseph Fourier and still preferred in some applications, such as signal processing.

Another way to define the Fourier sine and cosine transform can be as follows (see [14]):

$$T(\omega) = \left\{ \begin{array}{c} F_s(\omega) \\ F_c(\omega) \end{array} \right\} = \mathcal{T}\{\phi(y); x\} = \sqrt{\frac{2}{\pi}} \int_0^{\infty} \phi(y) \left\{ \begin{array}{c} \sin \omega x \\ \cos \omega x \end{array} \right\} \mathrm{d}x. \tag{10}$$

The Fourier sine and cosine transforms by Eq. (10) are self-inverse.

1.1.2 The Mellin transform

If an integrable function $f(x)$ on $(0, \infty)$ is $O(x^{-a})$ as $x \to 0+$ and $O(x^{-b})$, where $b > 1$, as $x \to \infty$, then the Mellin transform of $f(x)$ (see [15])

$$\varphi(s) = \mathcal{M}\{\phi(x); s\} = \int_0^\infty x^{s-1}\phi(x)\,\mathrm{d}x, \tag{11}$$

and its inverse transform of a function $\varphi(s)$

$$\phi(x) = \mathcal{M}^{-1}\{\varphi(s); x\} = \frac{1}{2\pi i}\int_{c-i\infty}^{c+i\infty} x^{-s}\varphi(s)\,\mathrm{d}s, \quad a < c < b. \tag{12}$$

is defined in the strip $a <$ Re $s < b$ (see [16]).

As far as the product of two functions goes, there holds (see [17])

$$\begin{aligned}\mathcal{M}\{\phi(x)g(x); s\} &= \int_0^\infty x^{s-1}\phi(x)g(x)\,\mathrm{d}x = \int_0^\infty x^{s-1}\phi(x)\mathcal{M}^{-1}\{\rho(s); x\}\,\mathrm{d}x \\ &= \frac{1}{2\pi i}\int_0^\infty x^{s-1}\phi(x)\int_{c-i\infty}^{c+i\infty} x^{-t}\rho(t)\,\mathrm{d}t\,\mathrm{d}x \\ &= \frac{1}{2\pi i}\int_{c-i\infty}^{c+i\infty} \rho(t)\,\mathrm{d}t\int_0^\infty x^{s-t-1}\phi(x)\,\mathrm{d}x \\ &= \frac{1}{2\pi i}\int_{c-i\infty}^{c+i\infty} \varphi(s-t)\rho(t)\,\mathrm{d}t \quad (\rho(s) = \mathcal{M}\{g(x); x\}),\end{aligned} \tag{13}$$

whereby the interchange of integrals is justified based on uniform convergence. So, Eq. (13) tells us that the Meliin transform of the product of two functions is the convolution of the individual Meliin transforms.

As regards the Fourier transform Eq. (8), we have

$$\mathcal{F}\{\phi(e^t); -is\} = \int_{-\infty}^{\infty} \phi(e^t)e^{it(-is)}\,\mathrm{d}t = \int_{-\infty}^{\infty} \phi(e^t)e^{ts}\,\mathrm{d}t. \tag{14}$$

After introducing the substitution $x = e^t$ in Eq. (14), the latter integral becomes

$$\int_0^\infty \phi(x)x^s\frac{\mathrm{d}x}{x} = \int_0^\infty \phi(x)t^{s-1}\,\mathrm{d}x = \mathcal{M}\{\phi(x); s\}. \tag{15}$$

So, the formulas Eq. (14) and Eq. (15) connect the Fourier and Mellin transform.

The gamma function is actually the Mellin transform of e^{-z}, that is,

$$\Gamma(z) = \int_0^\infty x^{z-1}e^{-z}\,\mathrm{d}x = \mathcal{M}\{e^{-z}; s\}, \quad \text{Re } z > 0, \tag{16}$$

and e^{-z} in Eq. (16) is expressed through the Mellin inverse transform

$$e^{-z} = \frac{1}{2\pi i}\int_{c-i\infty}^{c+i\infty} \Gamma(s)z^{-s}\,\mathrm{d}s, \quad |\arg z| < \frac{\pi}{2}. \tag{17}$$

1.1.3 The Hankel transform

Hankel transforms, named after Hermann Hankel (1839–1873), a German mathematician, are integral transforms whose kernels are Bessel or related functions, being the reason they are sometimes referred to as Bessel transforms (see [4]). Originally, the Hankel transform of order ν of a function $f(r)$ was defined by (see [13, 18])

$$F_\nu(s) = \mathcal{H}_\nu\{\phi(r); s\} = \int_0^\infty r\phi(r) J_\nu(sr)\, \mathrm{d}r, \tag{18}$$

where J_ν is the Bessel function of the first kind of order ν with $\nu > -\frac{1}{2}$. The famous Hankel integral formula gives the inversion formula

$$\phi(r) = \mathcal{H}_\nu^{-1}\{F_\nu(s); r\} = \int_0^\infty sJ_\nu(sr)\, \mathrm{d}s \int_0^\infty r\phi(r) J_\nu(sr)\, \mathrm{d}r. \tag{19}$$

Sufficient but not necessary conditions for the validity of Eq. (18) and Eq. (19) are as follows:

1. $\phi(r) = O\left(r^{-k}\right), r \to \infty$ where $k > \frac{3}{2}$.

2. $\phi'(r)$ is piecewise continuous over each bounded subinterval of $[0, \infty)$.

3. $\phi(r)$ is defined as $\frac{1}{2}(\phi(r+) + \phi(r-))$.

Inverting a Hankel transform of a function $f(r)$ is valid at every point at which $f(r)$, defined in $(0, \infty)$, is continuous, is piecewise continuous, and of bounded variation in every finite subinterval in $(0, \infty)$, and

4. $\int_0^\infty |f(r)| \sqrt{r}\, \mathrm{d}r < \infty$.

An alternative definition says that the Hankel transform of $g(r)$ is (see [19, p. 189])

$$G_\nu(s) = \mathcal{H}_\nu\{g(r); s\} = \int_0^\infty g(r) J_\nu(sr) \sqrt{sr}\, \mathrm{d}r. \tag{20}$$

The definitions Eq. (18) and Eq. (20) are related, since, if $g(r) = f(r)\sqrt{r}$, then $G_\nu(s) = F_\nu(s)\sqrt{s}$. It means the Hankel transform defined this way is self-inverse

$$g(r) = \mathcal{H}_\nu^{-1}\{G_\nu(s); r\} = \int_0^\infty G_\nu(s) J_\nu(sr) \sqrt{sr}\, \mathrm{d}s. \tag{21}$$

The obvious domain now has the condition $\int_0^\infty |g(r)|\, \mathrm{d}r < \infty$, but this can be extended. In the literature, some authors (see [4, 6]) use Eq. (20), that is, Eq. (21), as alternative definitions of the Hankel integral transform.

We shall call all the integral transforms having as the kernel Bessel or related functions the Hankel-type transforms. For instance, if we deal with Bessel or Struve functions, we replace φ_ν with J_ν or $\mathbf{H}_\nu$, and $\mathcal{H}_\nu\{\phi(y); x\}$ is $B_\nu(x)$ or $S_\nu(x)$, respectively, that is,

$$\left\{ \begin{array}{l} B_\nu(x) \\ S_\nu(x) \end{array} \right\} = \mathcal{H}_\nu\{\phi(y); x\} = \int_0^\infty y\phi(y) \left\{ \begin{array}{l} J_\nu(xy) \\ \mathbf{H}_\nu(xy) \end{array} \right\} \mathrm{d}y, \tag{22}$$

whereby in Eq. (22), we applied the original definition Eq. (18) of the Hankel transform.

1.2 Schlömilch series obtained from numerical series

We start with the partial fraction decomposition

$$\frac{p}{n^{2m}(n^2+p^2)}-\frac{(-1)^m}{p^{2m-1}(n^2+p^2)}=\sum_{k=1}^{m}\frac{(-1)^{m-k}}{n^{2k}p^{2m-2k+1}}, \tag{23}$$

which is easily proved by the mathematical induction method in m, distinguishing the cases $m=2q$ and $m=2q+1$, and the proof amounts to factoring the expressions $p^{4q}-n^{4q}$ or $p^{4q+2}+n^{4q+2}$.

From the identity Eq. (23), there follows

$$\sum_{n=1}^{\infty}\frac{(-1)^{n-1}p}{n^{2m}(n^2+p^2)}=\frac{(-1)^m}{p^{2m-1}}\sum_{n=1}^{\infty}\frac{(-1)^{n-1}}{n^2+p^2}+\sum_{k=1}^{m}\frac{(-1)^{m-k}}{p^{2m-2k+1}}\sum_{n=1}^{\infty}\frac{(-1)^{n-1}}{n^{2k}}. \tag{24}$$

Relying on the summation formula ([20, p. 685, 6])

$$\sum_{n=1}^{\infty}\frac{(-1)^{n-1}}{p^2+n^2}=\frac{1}{2p^2}-\frac{\pi}{2p}\operatorname{csch}(\pi p), \tag{25}$$

from Eq. (24), we arrive at the summation formula

$$\sum_{n=1}^{\infty}\frac{(-1)^{n-1}p}{n^{2m}(p^2+n^2)}=\frac{(-1)^{m+1}\pi}{2p^{2m}}\operatorname{csch}(\pi p)+\sum_{k=0}^{m}\frac{(-1)^k\eta(2m-2k)}{p^{2k+1}}, \quad \operatorname{Re} p>0, \tag{26}$$

where η presents Dirichlet's eta function $\eta(s)=\sum_{n=1}^{\infty}\frac{(-1)^{n-1}}{n^s}$.

Now, we can apply the inverse Mellin transform Eq. (12) to both sides of Eq. (26), knowing that (see [5, p. 342, Eq. (12) and p. 343, Eq. (16)])

$$\mathcal{M}^{-1}\left\{\frac{p}{p^2+n^2};x\right\}=\begin{cases}\cos(n\log x), & 0<x<1\\ 0, & x>1\end{cases}, \quad \operatorname{Re} p>0, \tag{27}$$

and

$$\mathcal{M}^{-1}\left\{\frac{1}{p^{\nu}};\right\}=\begin{cases}\dfrac{(-\log x)^{\nu-1}}{\Gamma(\nu)}, & 0<x<1\\ 0, & x>1\end{cases}, \quad \operatorname{Re}\ \nu>0, \quad \operatorname{Re} p>0. \tag{28}$$

From Eqs. (27) and (28), we obtain the sum of the series

$$\sum_{n=1}^{\infty}\frac{\cos(n\log x)}{n^{2m}}=\sum_{k=0}^{m}\frac{(-1)^k\eta(2m-2k)}{(2k)!}\log^{2k}x.$$

On the other hand, applying the Hankel transform Eq. (20) to the series Eq. (26), recalling that (see [6, p. 22, Eq. (7) and p. 24, Eq. (20)])

$$\mathcal{H}\left\{\left\{p^{\nu+\frac{1}{2}}(p^2+n^2)^{-\mu-1}\right\}(x)=\frac{n^{\nu-\mu}x^{\mu+\frac{1}{2}}}{2^{\mu}\Gamma(\mu+1)}K_{\nu-\mu}(nx), \quad n\in\mathbb{N}, \tag{29}$$

where K_ν, $-1 < \text{Re}\ \nu < 2\,\text{Re}\ \mu + \frac{3}{2}$, is the modified Bessel function of the second kind, and

$$\mathcal{H}\{\{p^\mu\}(x) = \frac{2^{\mu+\frac{1}{2}}}{x^{\mu+1}} \frac{\Gamma\left(\frac{\nu+\mu}{2} + \frac{3}{4}\right)}{\Gamma\left(\frac{\nu-\mu}{2} + \frac{1}{4}\right)}, \qquad -\text{Re}\,\nu - \frac{3}{2} < \text{Re}\ \mu < -\frac{1}{2}, \tag{30}$$

and calculating

$$\mathcal{H}\{\{\text{csch}(\pi p)\} = \int_0^\infty \sqrt{px} J_\nu(px)\text{csch}(\pi p)dp = \sqrt{\frac{1}{2\pi x}}\text{th}\frac{x}{2}, \quad |x| < \pi. \tag{31}$$

We conclude that we have to take $\nu = \frac{1}{2}$, so $\mu = 0$ in Eq. (29) and, because of the condition $-\text{Re}\ \nu - \frac{3}{2} < \text{Re}\ \mu < -\frac{1}{2}$, there must be $\mu = -1$ in Eq. (30). As a result, it is necessary to set $m = 0$ in Eq. (26), and taking account of Eq. (31), one obtains

$$\sum_{n=1}^\infty (-1)^{n-1}\sqrt{n}K_{1/2}(nx) = -\frac{1}{2}\sqrt{\frac{\pi}{2x}}\text{th}\frac{x}{2} + \frac{1}{2}\sqrt{\frac{\pi}{2x}} = \frac{1}{e^x+1}\sqrt{\frac{\pi}{2x}}, \quad |x| < \pi. \tag{32}$$

So, Eq. (32) presents the sum of the Schlömilch series over the modified Bessel functions of the second kind $K_{1/2}(x)$.

2. Series based on the Fourier sine and cosine transforms

An advantage of the proposed method for finding a summation formula of Schlömilch-type series involving integral transforms lies in the dispensability of their prior calculation; however, we can also obtain a new series by applying integral transforms. First, by making use of the Fourier sine and cosine transforms Eq. (9), p. 3, we form the series (see [21])

$$S_\alpha^T = \sum_{n=1}^\infty \frac{(s)^{n-1} T((an-b)x)}{(an-b)^\alpha} \tag{33}$$

where $a = \begin{Bmatrix} 1 \\ 2 \end{Bmatrix}$ $b = \begin{Bmatrix} 0 \\ 1 \end{Bmatrix}$, $s = 1$, or $s = -1$.
Theorem 1.1 The summation formula for the series Eq. (33) is

$$S_\alpha^T = \frac{c\pi x^{\alpha-1}\varphi(\alpha)}{2\Gamma(\alpha) f\left(\frac{\pi\alpha}{2}\right)} + \sum_{k=0}^\infty \frac{(-1)^k F(\alpha - 2k - \delta)x^{2k+\delta}}{(2k+\delta)!}\varphi(2k+\delta+1), \tag{34}$$

where, according to Eq. (11), we have applied the Mellin transform

$$\begin{aligned} \varphi(\alpha) &= \mathcal{M}\{\phi(y);\alpha\} = \int_0^\infty x^{\alpha-1}\phi(y)\,\mathrm{d}y, \\ \varphi(2k+\delta+1) &= \mathcal{M}\{\phi(y); 2k+\delta+1\} = \int_0^\infty x^{(2k+\delta+1)-1}\phi(y)\,\mathrm{d}y. \end{aligned} \tag{35}$$

The designation F in Eq. (34) stands for Riemann's zeta function or Dirichlet's function: eta, lambda, and beta. We obtain all particular cases by reading the other parameters from **Table 1**.

Proof. The method for obtaining summation formulas for Eq. (33) is based on replacing $T((an-b)x)$ with the right-hand side of Eq. (9) without previous calculation. As a result, the function sine or cosine, which we designate f, will appear in the numerator. In addition, we assume the function $|\phi(y)|$ is integrable. Because of that and the uniform convergence of the trigonometric series based on Dirichlet's test (see [22], p. 347) with respect to $y \in (0, \infty)$, we may interchange the order of summation and integration

$$\sum_{n=1}^{\infty} \frac{(s)^{n-1}}{(an-b)^{\alpha}} \int_0^{\infty} \phi(y) f((an-b)xy)\,dy = \int_0^{\infty} \phi(y) \sum_{n=1}^{\infty} \frac{(s)^{n-1} f((an-b)xy)}{(an-b)^{\alpha}}\,dy, \qquad (36)$$

applying afterward the formulas representing the trigonometric series as series in terms of Riemann zeta and related functions (see [23]), with xy instead of x, we find Eq. (34). Then, there remains still to apply the Mellin transforms Eq. (35) for a specific function ϕ. □

Concerning the convergence regions of the series Eq. (34), they will be the same as those in **Table 1** because the four convergence regions of the right-hand side trigonometric series Eq. (36) are $0 < xy < 2\pi$, $0 < |xy| < \pi$, $0 < xy < \pi$, $0 < |xy| < \frac{\pi}{2}$, and in view of the parameter y being $y \in (0, \infty)$, we have

$$0 < y < \frac{2\pi}{x}, \quad 0 < y < \frac{\pi}{|x|}, \quad 0 < y < \frac{\pi}{x}, \quad 0 < y < \frac{\pi}{2|x|}, \qquad (37)$$

so only x satisfying one of the conditions $0 < x < 2\pi$, $|x| < \pi$, $0 < x < \pi$, $|x| < \frac{\pi}{2}$ ensures that $0 < y < \infty$ in Eq. (37) or else, for values of x beyond these regions respectively, one would have the restriction $0 < y < 1$.

In some cases, listed in **Table 1**, when the right-hand side series truncates due to the vanishing of F functions, representation Eq. (34) takes on the closed form

a	b	s	c	F	f	δ	p	**Convergence region**
1	0				sin	1	0	
		1	1	ζ	cos	0	1	$0 < x < 2\pi$
					sin	1	0	
		−1	0	η	cos	0	1	$-\pi < x < \pi$
2	1				sin	1	0	
		1	$\frac{1}{2}$	λ	cos	0	1	$0 < x < \pi$
					sin	1	1	
		−1	0	β	cos	0	0	$-\frac{\pi}{2} < x < \frac{\pi}{2}$

Table 1.
General formula and closed-form cases, $\alpha = 2m + p - 1$.

$$S_{2m+p-1}^{T} = \frac{c\pi x^{2m+p-2}\varphi(2m+p-1)}{2(2m+p-2)!f\left(m\pi + \frac{\pi}{2}(p-1)\right)} + \sum_{k=0}^{m+p-1} \frac{(-1)^k F(2m-2k+p-1-\delta)x^{2k+\delta}}{(2k+\delta)!}\varphi(2k+\delta+1), \tag{38}$$

where $p = 0$ or $p = 1$ and $m \in \mathbb{N}$.

Example 2.1 If we choose $a = 1, b = 0, s = 1$ in Eq. (33), from **Table 1**, there follows $c = 1, F = \zeta$, and taking $f = \sin$, there must be $\alpha = 2m - 1$, that is, $p = 0$, $T = S$, and $\delta = 1$. For this choice of parameters, the series Eq. (38) becomes

$$S_{2m-1}^{S} = \sum_{n=1}^{\infty} \frac{S(nx)}{n^{2m-1}} = (-1)^{m-1} \frac{\pi x^{2m-2}}{2(2m-2)!}\varphi(2m-1) + \sum_{k=0}^{m-1} \frac{(-1)^k \zeta(2m-2k-2)x^{2k+1}}{(2k+1)!}\varphi(2k+2). \tag{39}$$

Consider the function $\phi(y)$ defined as $y(1-y^2)^{\nu-\frac{1}{2}}$ for $0 < y < 1$ and $\phi(y) = 0$ for $y > 1$, where Re $\nu > -\frac{1}{2}$, (see [24, p. 69, entry 7]), where we take $a = 1$. Applying the Fourier sine transform Eq. (9), p. 3, by virtue of ([2, p. 192, entry 11.5.1]), we find

$$S(x) = \int_0^{\infty} (1-y^2)^{\nu-\frac{1}{2}} \sin xy\, dy = \int_0^{1} (1-y^2)^{\nu-\frac{1}{2}} \sin xy\, dy = \frac{\sqrt{\pi}\Gamma\left(\nu+\frac{1}{2}\right)\mathbf{H}_\nu(x)}{2\left(\frac{x}{2}\right)^\nu}, \tag{40}$$

where $\mathbf{H}_\nu(x)$ denotes Struve's function (see [25, entry 11.2.1]), and by virtue of (see [26]), the right-hand side presents $s_{\nu,\nu}(x)/x^\nu$ with Lommel functions $s_{\mu,\nu}(x)$ (see [27, entries 11.9.3 & 11.9.4]). Thus, we have

$$\sum_{n=1}^{\infty} \frac{s_{\nu,\nu}(nx)}{n^{\nu+2m-1}} = (-1)^{m-1} \frac{\pi x^{\nu+2m-2}}{2(2m-2)!}\varphi(2m-1) + \sum_{k=0}^{m-1} \frac{(-1)^k \zeta(2m-2k-2)x^{\nu+2k+1}}{(2k+1)!}\varphi(2k+2), \tag{41}$$

and to determine $\varphi(2m-1)$ and $\varphi(2k+2)$ in Eq. (41), we read from (see [24, p. 311, entry 28]), where we set $h = 2$ and $\nu + \frac{1}{2}$ instead of ν

$$\begin{aligned} \varphi(2m-1) &= \mathcal{M}\{g(y); 2m-1\} = \frac{1}{2}\mathrm{B}\left(\nu+\frac{1}{2}, m-\frac{1}{2}\right), \quad \nu > -\frac{1}{2}, \\ \varphi(2k+2) &= \mathcal{M}\{g(y); 2k+2\} = \frac{1}{2}\mathrm{B}\left(\nu+\frac{1}{2}, k+1\right). \end{aligned} \tag{42}$$

Using Eq. (42), we find

$$\sum_{n=1}^{\infty} \frac{s_{\nu,\nu}(nx)}{n^{\nu+2m-1}} = (-1)^{m-1} \frac{\pi x^{\nu+2m-2}\Gamma\left(\nu+\frac{1}{2}\right)\Gamma\left(m-\frac{1}{2}\right)}{4(2m-2)!\Gamma(\nu+m)} + \sum_{k=0}^{m-1} \frac{(-1)^k \zeta(2m-2k-2)x^{\nu+2k+1}\Gamma\left(\nu+\frac{1}{2}\right)}{2^{k+1}(2k+1)!!\Gamma\left(\nu+k+\frac{3}{2}\right)}. \tag{43}$$

By Eq. (43), we obtain a finite sum of the series over Lommel functions.

2.1 A more general case

We are dealing with the series containing one more parameter and

$$S^T_{2m+p-1,\omega} = \sum_{n=1}^{\infty} \frac{(s)^{n-1} T((an-b)x)}{(an-b)^{2m+p-1}\left((an-b)^2-\omega^2\right)}, \tag{44}$$

where $\omega \in \mathbb{R}, |\omega| \neq an-b, m \in \mathbb{N}, p=0$ or $p=1$, $a = \left\{ \begin{matrix} 1 \\ 2 \end{matrix} \right\}$ $b = \left\{ \begin{matrix} 0 \\ 1 \end{matrix} \right\}$, $s=1$, or $s=-1$. After rearrangements, as with the series Eq. (33), we place the right-hand side of Eq. (9) instead of $T((an-b)x)$ in Eq. (44).

Theorem 1.2 The summation formula for the series Eq. (44) is

$$\begin{aligned} S^T_{2m+p-1,\omega} = \frac{1}{\omega^{2m}} \Bigg(& \frac{sp(1-b)}{2\omega^2} C(0) - \frac{s\pi \sin^b \frac{\pi\omega}{2}}{2\omega^p \sin \pi\omega} \Big(\cos \frac{\pi(s+1)(b+\omega)}{2a} T(\omega x) \\ & \mp \sin \frac{\pi(s+1)(b+\omega)}{2a} \tilde{T}(\omega x)\Big) + \sum_{k=1}^{m} \omega^{2k-2} \Bigg(\frac{(-1)^{k+p} c\pi x^{2k+p-2}}{2(2k+p-2)!} \varphi(2k+p-1) \\ & - \sum_{j=0}^{k+p-1} \frac{(-1)^j F(2k+p-1-2j-\delta) x^{2j+\delta}}{(2j+\delta)!} \varphi(2j+\delta+1) \Bigg) \Bigg), \end{aligned} \tag{45}$$

where $\omega \in \mathbb{R}, |\omega| \neq an-b, T(x) = \left\{ \begin{matrix} S(x) \\ C(x) \end{matrix} \right\}$, $\tilde{T}(x) = \left\{ \begin{matrix} C(x) \\ S(x) \end{matrix} \right\}$, $S(x)$, and $C(x)$ are the Fourier sine and cosine transforms of the function $g(y)$, and $\varphi(2k+p-1)$ and $\varphi(2j+\delta+1)$ present the Mellin transform

$$\begin{aligned} \varphi(2k+p-1) &= \mathcal{M}\{g(y); 2k+p-1\}, \\ \varphi(2k+\delta+1) &= \mathcal{M}\{g(y); 2k+\delta+1\}. \end{aligned} \tag{46}$$

The other relevant parameters are in **Table 1**, p. 8.

Proof. We rewrite Eq. (44), acting in the same manner as in the case of the trigonometric series [29], replacing f with T, that is,

$$\begin{aligned} S^T_{2m+p-1,\omega} = \frac{1}{\omega^{2m}} \Bigg(& \sum_{n=1}^{\infty} \frac{(s)^{n-1} T((an-b)x)}{(an-b)^{p-1}\left((an-b)^2-\omega^2\right)} \\ & - \sum_{k=1}^{m} \omega^{2k-2} \sum_{n=1}^{\infty} \frac{(s)^{n-1} T((an-b)x)}{(an-b)^{2k+p-1}} \Bigg), \end{aligned} \tag{47}$$

and begin with the first series in Eq. (47). Setting $m=0$ and using the Fourier sine or cosine transform $T((an-b)x)$, p. 3 again, we find

$$\begin{aligned} S^T_{p-1,\omega} &= \sum_{n=1}^{\infty} \frac{(s)^{n-1}}{(an-b)^{p-1}\left((an-b)^2-\omega^2\right)} \int_0^{\infty} \phi(y) f((an-b)xy)\, dy \\ &= \int_0^{\infty} \left(\sum_{n=1}^{\infty} \frac{(s)^{n-1} f((an-b)xy)}{(an-b)^{p-1}\left((an-b)^2-\omega^2\right)} \right) \phi(y)\, dy, \end{aligned} \tag{48}$$

where $f = \sin$ or $f = \cos$. So, relying on the summation formula for the trigonometric series Eq. (48) (see [28]), we find

$$S^T_{p-1,\omega} = \frac{sp(1-b)}{2\omega^2}\int_0^\infty \phi(y)\,\mathrm{d}y - \frac{s\pi\sin^b\frac{\pi\omega}{2}}{2\omega^p\sin\pi\omega}\int_0^\infty f_\omega(xy)\phi(y)\,\mathrm{d}y, \tag{49}$$

where for brevity, we have introduced $f_\omega(xy) = f\left(\omega xy - \frac{\pi}{2a}(s+1)(b+\omega)\right)$. Further, we apply the well-known trigonometric formulas to the right-hand side trigonometric function f, which we can rewrite in the form of

$$f(\alpha-\beta) = f(\alpha)\cos\beta \mp \tilde{f}(\alpha)\sin\beta, \tag{50}$$

with $f = \left\{ \begin{matrix} \sin \\ \cos \end{matrix} \right\} \tilde{f} = \left\{ \begin{matrix} \cos \\ \sin \end{matrix} \right\}$. Treating the first integral in Eq. (49) as obtained from the Fourier cosine transform $C(x)$ in Eq. (9), p. 3, by setting $x = 0$, and making use of Eq. (50), we get the following formula

$$\begin{aligned} S^T_{p-1,\omega} = \sum_{n=1}^\infty \frac{(s)^{n-1}T((an-b)x)}{(an-b)^{p-1}\left((an-b)^2-\omega^2\right)} = \frac{sp(1-b)}{2\omega^2}C(0) \\ -\frac{s\pi\sin^b\frac{\pi\omega}{2}}{2\omega^p\sin\pi\omega}\left(\cos\frac{\pi(s+1)(b+\omega)}{2a}T(\omega x)\mp\sin\frac{\pi(s+1)(b+\omega)}{2a}\tilde{T}(\omega x)\right), \end{aligned} \tag{51}$$

where $p = 0$ or $p = 1$, and $T(\omega x)$ and $\tilde{T}(\omega x)$ are the Fourier sine or cosine transforms Eq. (9) corresponding to $f(\omega x)$ and $\tilde{f}(\omega x)$, respectively.

There remains to deal with the second sum in Eq. (47). So, we replace again $T((an-b)x)$ with the Fourier sine or cosine transform of a function $g(y)$

$$\sum_{n=1}^\infty \frac{(s)^{n-1}T((an-b)x)}{(an-b)^{2k+p-1}} = \sum_{n=1}^\infty \frac{(s)^{n-1}}{(an-b)^{2k+p-1}}\int_0^\infty g(y)f((an-b)xy)\,\mathrm{d}y, \tag{52}$$

and from (52), we have

$$\sum_{n=1}^\infty \frac{(s)^{n-1}T((an-b)x)}{(an-b)^{2k+p-1}} = \int_0^\infty \sum_{n=1}^\infty \frac{(s)^{n-1}f((an-b)xy)}{(an-b)^{2k+p-1}}g(y)\,\mathrm{d}y. \tag{53}$$

We replace the right-hand side series in (53) with the closed-form trigonometric series [30, p. 446], and after applying the Mellin transform, we obtain

$$\begin{aligned} \sum_{n=1}^\infty \frac{(s)^{n-1}T((an-b)x)}{(an-b)^{2k+p-1}} = \frac{c\pi x^{2k+p-2}\varphi(2k+p-1)}{2(2k+p-2)!f\left(k\pi+\frac{\pi}{2}(p-1)\right)} \\ + \sum_{j=0}^{k+p-1}\frac{(-1)^k x^{2j+\delta}F(2k+p-1-2j-\delta)}{(2j+\delta)!}\varphi(2j+\delta+1). \end{aligned} \tag{54}$$

By combining Eq. (47) with Eqs. (51) and (54), we obtain Eq. (45). □

Example 2.2 Choosing $\phi(y) = (1 - y^2)^{\nu - \frac{1}{2}}$ for $0 \leqslant y < 1$ and $\phi(y) = 0$ and for $y \geqslant 1$ in Eq. (9), setting $\sin xy$ and $\cos xy$, respectively, then using correspondingly [2, p. 292, entry 11.5.1] and [2, p. 224, entry 10.9.4], we obtain

$$S(x) = \int_0^\infty (1 - y^2)^{\nu - \frac{1}{2}} \sin xy \, dy = \int_0^1 (1 - y^2)^{\nu - \frac{1}{2}} \sin xy \, dy = \frac{\sqrt{\pi}\Gamma\left(\nu + \frac{1}{2}\right)\mathbf{H}_\nu(x)}{2\left(\frac{x}{2}\right)^\nu},$$
$$C(x) = \int_0^\infty (1 - y^2)^{\nu - \frac{1}{2}} \cos xy \, dy = \int_0^1 (1 - y^2)^{\nu - \frac{1}{2}} \cos xy \, dy = \frac{\sqrt{\pi}\Gamma\left(\nu + \frac{1}{2}\right)J_\nu(x)}{2\left(\frac{x}{2}\right)^\nu}. \tag{55}$$

By including the above function in the Mellin transform, the formula Eq. (45) becomes

$$\sum_{n=1}^\infty \frac{S(nx)}{n^{2m-1}(n^2 - \omega^2)} = \frac{\pi\sqrt{\pi}\Gamma\left(\nu + \frac{1}{2}\right)}{4\left(\frac{x}{2}\right)^\nu \omega^{2m+\nu}} (J_\nu(\omega x) - \mathrm{ctg}(\pi\omega)\mathbf{H}_\nu(\omega x))$$
$$+ \sum_{k=1}^m \frac{\Gamma\left(\frac{1}{2} + \nu\right)}{\omega^{2m-2k+2}} \left(\frac{(-1)^k \pi x^{2k-2} \Gamma\left(k - \frac{1}{2}\right)}{4(2k-2)!\Gamma(k+\nu)} - \sum_{j=0}^{k-1} \frac{(-1)^j x^{2j+1} j! \zeta(2k - 2j - 2)}{2(2j+1)!\Gamma\left(\frac{3}{2} + j + \nu\right)} \right). \tag{56}$$

Replacing $S(nx)$ on the left-hand side of (56) with the corresponding expression in Eq. (55) and taking into account the explanation after Eq. (40), we arrive at the summation formula:

$$\sum_{n=1}^\infty \frac{s_{\nu,\nu}(nx)}{n^{2m-1}(n^2 - \omega^2)(nx)^\nu} = \frac{\pi}{2\omega^{2m}} \left(C(\omega x) - \mathrm{ctg}(\pi\omega) \frac{s_{\nu,\nu}(\omega x)}{(\omega x)^\nu} \right)$$
$$+ \sum_{k=1}^m \frac{\Gamma\left(\frac{1}{2} + \nu\right)}{\omega^{2m-2k+2}} \left(\frac{(-1)^k \pi x^{2k-2} \Gamma\left(k - \frac{1}{2}\right)}{4(2k-2)!\Gamma(k+\nu)} - \sum_{j=0}^{k-1} \frac{(-1)^j x^{2j+1} j! \zeta(2k - 2j - 2)}{2(2j+1)!\Gamma\left(\frac{3}{2} + j + \nu\right)} \right). \tag{57}$$

Thus, by Eq. (57), the series over the Lommel functions of the first kind have been represented as a finite sum.

Example 2.3 Now, we choose $\phi(y) = y(1 - y^2)^{\nu - \frac{1}{2}}$ for $0 \leqslant y \leqslant 1$ and $\phi(y) = 0$ for $y > 1$. In this case, we use [31, p. 437, entry 3.771.12.] and setting $u = 1$, we have

$$C(x) = \int_0^1 y(1 - y^2)^{\nu - \frac{1}{2}} \cos xy \, dx = \frac{1}{2\nu + 1} - \frac{\sqrt{\pi} 2^{\nu - 1} \Gamma\left(\nu + \frac{1}{2}\right)}{x^\nu} \mathbf{H}_{\nu+1}(x). \tag{58}$$

On the other hand, we place first $\nu + 1$ instead of ν in [27, entries 11.9.3 and 11.9.4], and then substituting $\nu - 1$ for μ, we establish the relation

$$
\begin{aligned}
-\frac{s_{\nu-1,\nu+1}(x)}{x^{\nu}} &= \frac{1}{2\nu+1}\sum_{m=0}^{\infty}\frac{(-1)^m\left(\frac{x}{2}\right)^{2m}\Gamma\left(\frac{1}{2}\right)\Gamma\left(\nu+\frac{3}{2}\right)}{\Gamma\left(m+\frac{1}{2}\right)\Gamma\left(m+\nu+\frac{3}{2}\right)} \\
&= \frac{1}{2\nu+1}-\frac{\sqrt{\pi}}{2}\Gamma\left(\nu+\frac{1}{2}\right)\sum_{m=0}^{\infty}\frac{(-1)^m\left(\frac{x}{2}\right)^{2m+2}}{\Gamma\left(m+\frac{3}{2}\right)\Gamma\left(m+\nu+\frac{5}{2}\right)} \\
&= \frac{1}{2\nu+1}-\frac{\sqrt{\pi}2^{\nu-1}\Gamma\left(\nu+\frac{1}{2}\right)}{x^{\nu}}\mathbf{H}_{\nu+1}(x).
\end{aligned}
\tag{59}
$$

Eq. (59) implies

$$
C(x)=\int_0^1 y\left(1-y^2\right)^{\nu-\frac{1}{2}}\cos xy\,\mathrm{d}x=-\frac{s_{\nu-1,\nu+1}(x)}{x^{\nu}}. \tag{60}
$$

Based on Eq. (60), setting $s=-1, T=C, a=2, b=1, m=0,$ and $p=0$ in Eq. (44), the series there reduces to the one over the Lommel functions, that is,

$$
\sum_{n=1}^{\infty}\frac{(-1)^{n-1}(2n-1)}{(2n-1)^2-\omega^2}C((2n-1)x)=\sum_{n=1}^{\infty}\frac{(-1)^n s_{\nu-1,\nu+1}((2n-1)x)}{x^{\nu}(2n-1)^{\nu-1}\left((2n-1)^2-\omega^2\right)}. \tag{61}
$$

We obtain the left-hand side series sum from Eq. (51) for $p=0$, and additionally, by virtue of Eq. (60), taking account of (61), we find

$$
\begin{aligned}
\sum_{n=1}^{\infty}\frac{(-1)^{n-1}s_{\nu-1,\nu+1}((2n-1)x)}{(2n-1)^{\nu-1}\left((2n-1)^2-\omega^2\right)} &= \frac{\pi x^{\nu}s_{\nu-1,\nu+1}(\omega x)}{4(\omega x)^{\nu}\cos\frac{\pi\omega}{2}} \\
&= \frac{\pi}{4\cos\frac{\omega\pi}{2}}\left(\frac{\sqrt{\pi}2^{\nu-1}\Gamma\left(\nu+\frac{1}{2}\right)}{\omega^{\nu}}\mathbf{H}_{\nu+1}(\omega x)-\frac{x^{\nu}}{2\nu+1}\right).
\end{aligned}
\tag{62}
$$

So, by Eq. (62), we come to a finite sum of another Lommel functions series.

3. Series based on the Hankel-type integral transforms

Now, we are going to obtain the sums of the series

$$
I_{\alpha}^{D}=\sum_{n=1}^{\infty}\frac{(s)^{n-1}D_{\nu}((an-b)x)}{(an-b)^{\alpha}},\quad \alpha>0,\quad \nu>-\frac{1}{2}, \tag{63}
$$

where $a=\left\{\begin{matrix}1\\2\end{matrix}\right\}$ $b=\left\{\begin{matrix}0\\1\end{matrix}\right\}$, $s=1$ or -1, and $D_{\nu}(x)=\left\{\begin{matrix}B_{\nu}(x)\\S_{\nu}(x)\end{matrix}\right\}$ $\varphi_{\nu}(x)=\left\{\begin{matrix}J_{\nu}(x)\\\mathbf{H}_{\nu}(x)\end{matrix}\right\}$, with B_{ν} and S_{ν} as the Hankel integrals Eq. (22), p. 6.

Theorem 1.3 The summation of the series Eq. (63) is

$$
\begin{aligned}
I_\alpha^D &= \frac{c\pi x^{\alpha-1}}{2^\alpha \Gamma\left(\frac{\alpha-\nu+1}{2}\right)\Gamma\left(\frac{\alpha+\nu+1}{2}\right) f\left(\frac{\pi(\alpha-\nu)}{2}\right)}\varphi(\alpha) \\
&+ \sum_{k=0}^{\infty} \frac{(-1)^k F(\alpha-\nu-2k-\delta) x^{\nu+2k+\delta}}{2^{\nu+2k+\delta}\Gamma\left(k+1+\frac{\delta}{2}\right)\Gamma\left(\nu+k+1+\frac{\delta}{2}\right)}\varphi(\nu+2k+\delta+1),
\end{aligned} \tag{64}
$$

where are $\varphi(\alpha)$ and $\varphi(\nu+2k+\delta+1)$ are the Mellin transforms

$$
\begin{aligned}
\varphi(\alpha) &= \int_0^\infty y^{\alpha-1}\phi(y)\,\mathrm{d}y, \\
\varphi(\nu+2k+\delta+1) &= \int_0^\infty \phi(y) y^{\nu+2k+\delta}\,\mathrm{d}y.
\end{aligned} \tag{65}
$$

Proof. Inserting in Eq. (63), the Hankel integral Eq. (22), setting previously there $(an-b)x$ instead of x, we have

$$
\begin{aligned}
I_\alpha^D &= \sum_{n=1}^{\infty} \frac{(s)^{n-1}}{(an-b)^\alpha}\int_0^\infty y\phi(y)\varphi_\nu((an-b)xy)\,\mathrm{d}y \\
&= \int_0^\infty \sum_{n=1}^{\infty} \frac{(s)^{n-1}\varphi_\nu((an-b)xy)}{(an-b)^\alpha} y\phi(y)\,\mathrm{d}y.
\end{aligned} \tag{66}
$$

where φ_ν is J_ν or $\mathbf{H}_\nu$. Replacing the right-hand series of Eq. (66) with the formula [28, p. 22, Eq. (12)], changing there x to xy, we get

$$
\begin{aligned}
I_\alpha^D = \int_0^\infty &\left(\frac{c\pi\left(\frac{xy}{2}\right)^{\alpha-1}}{2\Gamma\left(\frac{\alpha-\nu+1}{2}\right)\Gamma\left(\frac{\alpha+\nu+1}{2}\right) f\left(\frac{\pi(\alpha-\nu)}{2}\right)} \right. \\
&\left. + \sum_{k=0}^{\infty} \frac{(-1)^k \left(\frac{xy}{2}\right)^{\nu+2k+\delta} F(\alpha-\nu-2k-\delta)}{\Gamma\left(k+1+\frac{\delta}{2}\right)\Gamma\left(\nu+k+1+\frac{\delta}{2}\right)} \right) y\phi(y)\,\mathrm{d}y,
\end{aligned} \tag{67}
$$

where $\varphi_\nu = \begin{Bmatrix} J_\nu \\ \mathbf{H}_\nu \end{Bmatrix} \delta = \begin{Bmatrix} 0 \\ 1 \end{Bmatrix}$. From (67), we obtain

$$
\begin{aligned}
I_\alpha^D &= \frac{c\pi x^{\alpha-1}}{2^\alpha \Gamma\left(\frac{\alpha-\nu+1}{2}\right)\Gamma\left(\frac{\alpha+\nu+1}{2}\right) f\left(\frac{\pi(\alpha-\nu)}{2}\right)}\int_0^\infty y^{\alpha-1}\phi(y)\,\mathrm{d}y \\
&+ \sum_{k=0}^{\infty} \frac{(-1)^k F(\alpha-\nu-2k-\delta) x^{\nu+2k+\delta}}{2^{\nu+2k+\delta}\Gamma\left(i+1+\frac{\delta}{2}\right)\Gamma\left(\nu+k+1+\frac{\delta}{2}\right)}\int_0^\infty y^{\nu+2k+\delta}\phi(y)\,\mathrm{d}y.
\end{aligned} \tag{68}
$$

To calculate the integrals in (68), we apply the Mellin transforms of $\phi(y)$ given by Eq. (65). Thus, we arrive at Eq. (63). For other parameters, you can look up in **Table 1**, p. 8. □

The formula Eq. (64) takes a closed form in the same cases as the formula [28, p. 22, Eq. (12)].

Example 3.1 We choose in Eq. (22) $\phi(y) = y^{\nu}(1-y^2)^{-\frac{1}{2}}$ for $0<y<1$ and $\phi(y)=0$ for $y \geqslant 1$. Since (see [31, p. 680]), we have

$$B_{\nu}(x) = \mathcal{H}_{\nu}\{\phi(y);x\} = \int_0^{\infty} \frac{y}{y^{\nu}\sqrt{1-y^2}} J_{\nu}(xy)\,\mathrm{d}y = \sqrt{\frac{\pi}{2x}}\mathbf{H}_{\nu-\frac{1}{2}}(x). \tag{69}$$

So, we make use of Eq. (69) to deal with the series

$$I_{\alpha}^{B} = \sum_{n=1}^{\infty} \frac{(s)^{n-1} B_{\nu}(nx)}{n^{\alpha}} = \sqrt{\frac{\pi}{2x}} \sum_{n=1}^{\infty} \frac{(s)^{n-1}\mathbf{H}_{\nu-\frac{1}{2}}(nx)}{n^{\alpha+\frac{1}{2}}}. \tag{70}$$

Applying the general formula for the summation of the series over Bessel or Struve functions (see [28]), where we take $f = \sin$, $\alpha + \frac{1}{2}$, and $\nu - \frac{1}{2}$ instead of α and ν, respectively, for Eq. (70), we get

$$I_{\alpha}^{B} = \frac{c\pi\sqrt{\pi}\left(\frac{x}{2}\right)^{\alpha-1}}{4\Gamma\left(\frac{\alpha-\nu+2}{2}\right)\Gamma\left(\frac{\alpha+\nu+1}{2}\right)\cos\frac{\pi(\alpha-\nu)}{2}} + \sum_{k=0}^{\infty} \frac{(-1)^k\sqrt{\pi}F(\alpha-\nu-2k)\left(\frac{x}{2}\right)^{\nu+2k}}{2\Gamma\left(k+\frac{3}{2}\right)\Gamma(\nu+k+1)}, \tag{71}$$

and the parameters s, c, F in (71) we read from **Table 1**, p. 8.

3.1 A more general series

Now, we consider the series

$$I_{2m-\mu,\omega}^{D} = \sum_{n=1}^{\infty} \frac{(s)^{n-1} D_{\nu}((an-b)x)}{(an-b)^{2m-\mu}\left((an-b)^2-\omega^2\right)}, \quad \nu > -\frac{1}{2}, \tag{72}$$

where $\alpha = 2m-\mu, m\in\mathbb{N}, \omega\in\mathbb{R}, \ |\omega| \ \neq an-b$, and $\mu+\nu = 1-p$, with $p=0$ or $p=1$. The other parameters are the same as in Eq. (63).

We derive the formula for the summation of series Eq. (72), following the same train of thought as in the case of the series Eq. (108), so we prove the theorem.

Theorem 1.4. The summation formula for the series Eq. (72) is

$$\begin{aligned} I_{2m-\mu,\omega}^{D} = \frac{\left(\frac{x}{2}\right)^{\nu}}{\omega^{2m+2}} & \left(\frac{s(1-b)p}{2\Gamma(1+\nu)}\varphi(\nu+1) - \frac{s\pi\sin^{b}\frac{\pi\omega}{2}}{2\left(\frac{x}{2}\right)^{\nu}\omega^{\nu+p-2}\sin\pi\omega}\Omega_{\nu}(\omega x)\right. \\ & + \sum_{k=1}^{m}\omega^{2k}\left(\frac{(-1)^{k+p}c\sqrt{\pi}x^{2k-2+p}\Gamma\left(k+\frac{p-1}{2}\right)}{2(2k-2+p)!\Gamma\left(k+\nu+\frac{p}{2}\right)}\varphi(2k+\nu+p)\right. \\ & \left.\left. - \sum_{j=0}^{k-1+p}\frac{(-1)^j x^{2j+\delta}F(2k-1-2j+p-\delta)\Gamma\left(j+\frac{\delta+1}{2}\right)}{\sqrt{\pi}(2j+\delta)!\Gamma\left(\nu+j+1+\frac{\delta}{2}\right)}\varphi(2j+\nu+\delta+2)\right)\right), \end{aligned} \tag{73}$$

where $\Omega_\nu(\omega x) = D_\nu(\omega x)\cos\frac{\pi}{2a}(s+1)(b+\omega) - \tilde{D}_\nu(\omega x)\sin\frac{\pi}{2a}(s+1)(b+\omega)$, $D_\nu = \begin{Bmatrix} B_\nu \\ S_\nu \end{Bmatrix}$ $\tilde{D}_\nu = \begin{Bmatrix} -S_\nu \\ B_\nu \end{Bmatrix}$, with $\varphi(\nu+1)$, $\varphi(2k+\nu+p)$, and $\varphi(2j+\nu+\delta+2)$ as the Mellin transforms. The parameters $a, b, s, c,$ and F are in **Table 1**, p. 8.

Proof. We first replace $D_\nu((an-b)x)$ with the Hankel integral, and then we interchange summation and integration

$$I^D_{2m-\mu,\omega} = \sum_{n=1}^{\infty} \frac{(s)^{n-1}(an-b)^{\mu}}{(an-b)^{2m}\left((an-b)^2-\omega^2\right)} \int_0^\infty y\phi(y)\varphi_\nu((an-b)xy)\,dy$$
$$= \int_0^\infty y\phi(y) \sum_{n=1}^{\infty} \frac{(s)^{n-1}\varphi_\nu((an-b)xy)}{(an-b)^{2m-\mu}\left((an-b)^2-\omega^2\right)}\,dy, \tag{74}$$

and referring to the summation formula of a more general series over Bessel or Struve functions [23], on account of $\mu+\nu = 1-p$, we have

$$S^{\varphi}_{2m-\mu,\omega} = \frac{\left(\frac{x}{2}\right)^\nu}{\omega^{2m+2}} \frac{sp(1-b)y^\nu}{2\Gamma(1+\nu)} - \frac{s\pi \sin^b \frac{\pi\omega}{2}}{2\omega^{2m+\nu+p}\sin \pi\omega}\Phi_\nu(\omega xy)$$
$$+ \frac{\left(\frac{x}{2}\right)^\nu}{\omega^{2m+2}} \sum_{k=1}^{m} \omega^{2k} \left(\frac{(-1)^{k+p} c\sqrt{\pi} x^{2k-2+p}\Gamma\left(k+\frac{p-1}{2}\right)}{2(2k-2+p)!\Gamma\left(k+\nu+\frac{p}{2}\right)} y^{2k+\nu+p-2} \right. \tag{75}$$
$$\left. - \sum_{j=0}^{k-1+p} \frac{(-1)^j x^{2j+\delta} F(2k-1-2j+p-\delta)\Gamma\left(j+\frac{\delta+1}{2}\right)}{\sqrt{\pi}(2j+\delta)!\Gamma\left(\nu+j+1+\frac{\delta}{2}\right)} y^{2j+\nu+\delta} \right),$$

where $\Phi_\nu(\omega xy) = \varphi_\nu(\omega xy)\cos\frac{\pi}{2a}(s+1)(b+\omega) - \tilde{\varphi}_\nu(\omega xy)\sin\frac{\pi}{2a}(s+1)(b+\omega)$, $\varphi_\nu = \begin{Bmatrix} J_\nu \\ \mathbf{H}_\nu \end{Bmatrix}$ $\tilde{\varphi}_\nu = \begin{Bmatrix} -\mathbf{H}_\nu \\ J_\nu \end{Bmatrix}$. After substituting Eq. (75) for the right-hand side sum in Eq. (74) and then integrating, we obtain Eq. (73). □

Example 3.2 We choose $\phi(y) = y^{\nu-1}(1-y^2)^{\nu-\frac{1}{2}}$. In this case (see [31, p. 680]),

$$B_\nu(x) = \mathcal{H}_\nu\{\phi(y);x\} = \int_0^\infty y^\nu (1-y^2)^{\nu-\frac{1}{2}} J_\nu(xy)\,dy = \frac{\sqrt{\pi}\Gamma\left(\nu+\frac{1}{2}\right)J^2_\nu\left(\frac{x}{2}\right)}{2\left(\frac{x}{2}\right)^\nu}. \tag{76}$$

Replacing the rightmost expression of Eq. (76) in Eq. (72) and setting $a=1, b=0, s=-1, m=0$ and $p=1$, which implies $\mu+\nu=0$, we have

$$I^B_{-\mu,\omega} = \frac{\sqrt{\pi}\Gamma\left(\nu+\frac{1}{2}\right)}{2\left(\frac{x}{2}\right)^\nu} \sum_{n=1}^{\infty} \frac{(-1)^{n-1}J^2_\nu\left(\frac{nx}{2}\right)}{n^{2\nu}(n^2-\omega^2)}. \tag{77}$$

On the other hand, upon implementing Eq. (73), with $D_\nu = B_\nu$ and the above parameters, we establish the relation with Eq. (77)

$$I^{B}_{-\mu,\omega} = \frac{\sqrt{\pi}\Gamma\left(\nu+\frac{1}{2}\right)}{2\left(\frac{x}{2}\right)^{\nu}} \sum_{n=1}^{\infty} \frac{(-1)^{n-1} J_{\nu}^{2}\left(\frac{nx}{2}\right)}{n^{2\nu}(n^2-\omega^2)} = \frac{\pi\omega^{-\nu-1}}{2\sin\pi\omega} B_{\nu}(\omega x) - \frac{\left(\frac{x}{2}\right)^{\nu}\varphi(\nu+2)}{2\omega^2\Gamma(\nu+1)}. \tag{78}$$

Calculating the Mellin transform $\varphi(\nu+2)$, it yields

$$\varphi(\nu+2) = \int_0^1 y^{(\nu+2)-1}\phi(y)\,\mathrm{d}y = \int_0^1 y^{2\nu}\left(1-y^2\right)^{\nu-\frac{1}{2}}\mathrm{d}y = \frac{\sqrt{\pi}\Gamma\left(\nu+\frac{1}{2}\right)}{2^{2\nu+1}\Gamma(\nu+1)}. \tag{79}$$

Based on Eqs. (78) and (79), there holds

$$I^{B}_{-\mu,\omega} = \frac{\sqrt{\pi}\Gamma\left(\nu+\frac{1}{2}\right) J_{\nu}^{2}\left(\frac{\omega x}{2}\right)}{4\left(\frac{x}{2}\right)^{\nu}\omega^{2\nu+1}\sin\pi\omega} - \frac{\left(\frac{x}{2}\right)^{\nu}\sqrt{\pi}\Gamma\left(\nu+\frac{1}{2}\right)}{4^{\nu+1}\omega^2\Gamma^2(\nu+1)}. \tag{80}$$

As a collateral result, we obtain the sum of another summation formula involving the Bessel functions:

$$\sum_{n=1}^{\infty} \frac{(-1)^{n-1} J_{\nu}^{2}\left(\frac{nx}{2}\right)}{n^{2\nu}(n^2-\omega^2)} = \frac{\pi J_{\nu}^{2}\left(\frac{\omega x}{2}\right)}{2\omega^{2\nu+1}\sin\pi\omega} - \frac{\left(\frac{x}{2}\right)^{2\nu}}{2^{2\nu+1}\omega^2\Gamma^2(\nu+1)}. \tag{81}$$

The left-hand side series in (81) is represented by a finite sum.

4. Product of the Fourier sine or cosine integral and a trigonometric function

We shall consider here the series and their summation formulas

$$I^{T,g}_{\alpha} = \sum_{n=1}^{\infty} \frac{(s)^{n-1} T((an-b)x)}{(an-b)^{\alpha}} g((an-b)z), \quad \alpha \in \mathbb{R}^{+}, \tag{82}$$

where $a = \begin{Bmatrix} 1 \\ 2 \end{Bmatrix}$, $b = \begin{Bmatrix} 0 \\ 1 \end{Bmatrix}$, $s = 1$ or -1, the function g denotes sin or cos and $T(x) = \begin{Bmatrix} S(x) \\ C(x) \end{Bmatrix}$ $f(x) = \begin{Bmatrix} \sin x \\ \cos x \end{Bmatrix}$, with $T(x)$ as the Fourier sine or cosine transform Eq. (9), p. 3, i.e.

$$T(x) = \mathcal{T}\{\phi(y);x\} = \int_0^{\infty} \phi(y) f(xy)\,\mathrm{d}y. \tag{83}$$

To find a sum of the series Eq. (82), we replace $T((an-b)x)$ in the series Eq. (82) with the Fourier sine or cosine integral, then providing that interchange summation and integration is permissible [32], we come to the equality

$$\begin{aligned} I^{T,g}_{\alpha} &= \sum_{n=1}^{\infty} \frac{(s)^{n-1} g((an-b)z)}{(an-b)^{\alpha}} \int_0^{\infty} \phi(y) f((an-b)xy)\,\mathrm{d}y \\ &= \int_0^{\infty} \phi(y) \sum_{n=1}^{\infty} \frac{(s)^{n-1} f((an-b)xy) g((an-b)z)}{(an-b)^{\alpha}}\,\mathrm{d}y, \quad \alpha \in \mathbb{R}^{+}. \end{aligned} \tag{84}$$

The following lemma confirms the equality Eq. (84).

Lemma 4.1 For an absolutely integrable function $\phi(y)$ on $y \in (0, \infty)$, summation and integration in Eq. (84) are interchangeable.

Proof. Let $S_n(x, y, z)$ denote the nth partial sum of the trigonometric series $S(x, y, z)$ on the right-hand side of Eq. (84), which is, according to Dirichlet's test (see [22], p. 347), uniformly convergent with respect to $y \in (0, \infty)$ as n approaches infinity, since the partial sums $S_n(x, y, z)$ are uniformly bounded and the sequence $1/(an - b)^{\alpha}$ monotonically tends to zero.

Depending on the choice of parameters, there are four regions. In the case, for $a = 1, b = 0, s = 1$, we have $\{(x, y) | 0 < y|x| < z, y|x| < z < 2\pi - y|x|\}$, whence there easily follows

$$0 < y < \frac{z}{|x|}, \quad y < \frac{2\pi - z}{|x|}, \quad 0 < y < \frac{\pi}{|x|}. \tag{85}$$

It is through (x, z) belonging to $K_1 = \{(x, z),\ x| \ < \pi|, x| \ < z < 2\pi - |x||\}$ that the inequalities Eq. (85) ensure the interval $1 < y < \infty$. Obviously, $(x, z) \in K_1$ implies $0 < y \leqslant 1$ as well. Similarly, the other convergence regions are
$K_2 = \{(x, z), |x| < \pi, |x| - \pi < z < \pi - |x|\}$, for $a = 1, b = 0, s = -1$,
$K_3 = \{(x, z), |x| < \frac{\pi}{2}, |x| - \pi < z < \pi - |x|\}$, for $a = 2, b = 1, s = 1$, and
$K_4 = \{(x, z), |x| < \frac{\pi}{2}, |x| - \frac{\pi}{2} < z < \frac{\pi}{2} - |x|\}$, for $a = 2, b = 1, s = -1$.

Further, for each $M > 0$, we have

$$\begin{aligned} F_n(M, x, z) &= \sum_{k=1}^{n} \frac{(s)^{n-1} g((ak - b)z)}{(ak - b)^{\alpha}} \int_{1/M}^{M} \phi(y) f((ak - b)xy)\, dy \\ &= \int_{1/M}^{M} \phi(y) \sum_{k=1}^{n} \frac{(s)^{k-1} f((ak - b)xy) g((ak - b)z)}{(ak - b)^{\alpha}}\, dy. \end{aligned} \tag{86}$$

Because of uniform convergence $\lim\limits_{n \to \infty} S_n(x, y, z) = S(x, y, z)$, we may interchange the signs for limit and integral in Eq. (86)

$$\lim_{n \to \infty} F_n(M, x, z) = \int_{1/M}^{M} \phi(y) \lim_{n \to \infty} S_n(x, y, z)\, dy = \int_{1/M}^{M} \phi(y) S(x, y, z)\, dy. \tag{87}$$

Taking account of $|\phi(y)|$ being integrable and $S_n(x, y, z)$ uniformly bounded, there exists $\varepsilon > 0$ such that

$$\left| \int_0^{\infty} \phi(y) S_n(x, y, z)\, dy \right| \leqslant \int_0^{\infty} |\phi(y)| \cdot |S_n(x, y, z)|\, dy \leqslant \frac{1}{\sin \varepsilon} \int_0^{\infty} |\phi(y)|\, dy. \tag{88}$$

The improper integral on the left-hand side of the inequality Eq. (88) is uniformly convergent with respect to n, which implies

$$\lim_{M \to \infty} F_n(M, x, z) = \int_0^{\infty} \phi(y) S_n(x, y, z)\, dy. \tag{89}$$

In other words, this limit exists independently of n, which allows to swap the limits:

$$\lim_{n\to\infty}\lim_{M\to\infty}F_n(M,x,z)=\lim_{M\to\infty}\lim_{n\to\infty}F_n(M,x,z)=\int_0^\infty \phi(y)S(x,y,z)\,dy. \tag{90}$$

Applying Eqs. (90), (89) and (87) to Eq. (86) gives rise to Eq. (84), making there the interchange of integration and summation legitimate. □

Relying on the preceding lemma, we can determine a sum of the series Eq. (82), that is, there holds.

Theorem 1.5 The summation formula for the series Eq. (82) is

$$\begin{aligned} I_\alpha^{T,g} &= \frac{\tau c\pi}{2\Gamma(\alpha)h\left(\frac{\pi\alpha}{2}\right)}\sum_{k=0}^{\infty}\binom{\alpha-1}{2k+d}z^{\alpha-1-2k-d}x^{2k+d}\varphi(2k+d+1)\\ &+\tau\sum_{k=0}^{\infty}\frac{(-1)^k F(\alpha-2k-\delta)}{(2k+\delta)!}\sum_{j=0}^{k-v}\binom{2k+\delta}{2j+d}z^{2k+\delta-2j-d}x^{2j+d}\varphi(2j+d+1), \end{aligned} \tag{91}$$

where $f=\begin{Bmatrix}\sin\\ \cos\end{Bmatrix}d=\begin{Bmatrix}1\\0\end{Bmatrix}$, $\begin{Bmatrix}f=g\\ f\neq g\end{Bmatrix}\delta=\begin{Bmatrix}0\\1\end{Bmatrix}h=\begin{Bmatrix}\cos\\ \sin\end{Bmatrix}$, and $\tau=-1, v=1$ only if $f=g=\sin$, and otherwise, $\tau=1,\ \ v=0$, $\varphi(2k+d+1)$ and $\varphi(2j+d+1)$ present the Mellin transforms of $\phi(y)$, that is,

$$\varphi(2k+d+1)=\int_0^\infty y^{2k+d}\phi(y)\,dy,\quad \varphi(2j+d+1)=\int_0^\infty y^{2j+d}\phi(y)\,dy. \tag{92}$$

Proof. We rely on the summation formula for the product of two trigonometric functions [23] with z instead of y and xy instead of x so that we can use it as a sum for the right-hand side series in Eq. (84). Afterward, by applying integration and taking account of Mellin transform Eq. (92), we obtain Eq. (91). □

As for the convergence regions, we follow the same line of reasoning as that for the series Eq. (33), p. 8, and find that they are those on p. 16.

4.1 Closed-form cases

Closed-form cases of the formula Eq. (91) ensure if $\alpha=\begin{Bmatrix}2m\\ 2m-1\end{Bmatrix}h=\begin{Bmatrix}\cos\\ \sin\end{Bmatrix}$, so one obtains

$$\begin{aligned} I_{2m-r}^{T,g} &= \frac{\tau c\pi}{2}\sum_{k=0}^{m-r}\frac{(-1)^{m-r}z^{2m-r-1-2k-d}x^{2k+d}\varphi(2k+d+1)}{(2m-r-1-2k-d)!(2k+d)!}\\ &+\tau\sum_{k=0}^{m-r}(-1)^k F(2m-r-2k-\delta)\sum_{j=0}^{k-v}\frac{z^{2k+\delta-2j-d}x^{2j+d}\varphi(2j+d+1)}{(2k+\delta-2j-d)!(2k+\delta)!}, \end{aligned} \tag{93}$$

where $f=\begin{Bmatrix}\sin\\ \cos\end{Bmatrix}d=\begin{Bmatrix}1\\0\end{Bmatrix}$, $\begin{Bmatrix}f=g\\ f\neq g\end{Bmatrix}\delta=\begin{Bmatrix}0\\1\end{Bmatrix}, r=\begin{Bmatrix}0\\1\end{Bmatrix}h=\begin{Bmatrix}\cos\\ \sin\end{Bmatrix}$, and $\tau=-1, v=1$ only if $f=g=\sin$; otherwise, $\tau=1, v=0$. So, if we set $r=0$, there must be $h=\cos$, and for $r=1$, we have to take $h=\sin$.

Example 4.1 We choose $\phi(y)=(1+y)^{-1}$ for $0<y<1$ and $\phi(y)=0$ for $y>1$ in Eq. (9). From Eq. (93), for $a=1$, $b=0$, and $s=1$, we take $c=1$ and $F=\zeta$, and for

$f = \sin$ and $g = \cos$, there follows $\delta = 1$ and $d = 1$. Additionally, for $m = 2$, we choose $r = 1$, and take $h = \sin$, so we have

$$\begin{aligned} I_3^{S,\cos} &= \sum_{n=1}^{\infty} \frac{\cos nz}{n^3} \int_0^{\infty} \phi(y) \sin nxy \, dy = \sum_{n=1}^{\infty} \frac{\cos nz}{n^3} \int_0^1 \frac{\sin nxy}{1+y} dy \\ &= -\frac{\pi}{2} zx\varphi(2) + \sum_{k=0}^{1} \frac{(-1)^k \zeta(2-2k)}{(2k+1)!} \sum_{j=0}^{k} \binom{2k+1}{2j+1} z^{2k-2j} x^{2j+1} \varphi(2j+2), \end{aligned} \tag{94}$$

with $\varphi(s)$ as the Mellin transform of $f(x)$. In this case, we express it through the digamma function designated as ψ (see [33, p. 14, entry 2.7])

$$\varphi(s) = \frac{1}{2}\left(\psi\left(\frac{s+1}{2}\right) - \psi\left(\frac{s}{2}\right)\right), \quad \psi(s) = \frac{\Gamma'(s)}{\Gamma(s)}. \tag{95}$$

Taking account of (see [2, p. 137, entry 5.4.15])

$$\psi\left(n + \frac{1}{2}\right) = -\gamma - \log 4 + \sum_{k=1}^{n} \frac{2}{2k-1}, \tag{96}$$

from Eqs. (95) and (96), we find

$$\begin{aligned} \varphi(2) &= 1 - \log 2 \\ \varphi(2j+2) &= \frac{1}{2}\left(\psi\left(j + \frac{3}{2}\right) - \psi(j+1)\right) \\ &= \frac{1}{2}\left(H_{j+\frac{1}{2}} - H_j\right) = H_{2j+1} - H_j - \log 2, \end{aligned} \tag{97}$$

where $H_n = \gamma + \psi(n+1)$ denotes the nth harmonic number. So, using Eq. (97), for $I_3^{S,\cos}$ in Eq. (94), one obtains

$$I_3^{S,\cos} = \frac{x^3}{72}(5 - 6\log 2) - x\left(\frac{z^2}{4} - \frac{\pi z}{2} + \frac{\pi^2}{6}\right)(\log 2 - 1). \tag{98}$$

This way, we represent the series Eq. (91) for the specific parameters as a finite sum.

4.2 Limiting value cases

However, if $\alpha = \begin{Bmatrix} 2m \\ 2m-1 \end{Bmatrix}$ $h = \begin{Bmatrix} \sin \\ \cos \end{Bmatrix}$ in Eq. (91), one should take a limit. Choosing the same parameters as in the previous example, and $\alpha = 2m$, on the left-hand side, we have

$$I_{2m}^{S,\cos} = \sum_{n=1}^{\infty} \frac{\cos nz}{n^{2m}} \int_0^1 \frac{\sin nxy}{1+y} dy. \tag{99}$$

However, since $h = \sin$, we are not permitted to replace immediately α with $2m$, $m \in \mathbb{N}$, on the right-hand side of Eq. (91), we have to take a limit, but split beforehand both series in two, up to $k = m - 1$ and from $k = m$. Firstly, we are dealing with

$$\lim_{\alpha\to 2m}\left(\frac{\pi}{2\Gamma(\alpha)\sin\dfrac{\pi\alpha}{2}}\sum_{k=0}^{m-1}\binom{\alpha-1}{2k+1}z^{\alpha-2k-2}x^{2k+1}\varphi(2k+2)\right.$$
$$\left.+\sum_{k=0}^{m-1}\frac{(-1)^k\zeta(\alpha-2k-1)}{(2k+1)!}\sum_{j=0}^{k}\binom{2k+1}{2j+1}z^{2k-2j}x^{2j+1}\varphi(2j+2)\right),\tag{100}$$

and designating $\Phi_{2m}^{(1)}(x,z)$ the limiting value of (100), there holds

$$\Phi_{2m}^{(1)}(x,z)=\frac{(-1)^{m-1}}{(2m-1)!}\sum_{k=0}^{m-1}\binom{2m-1}{2k}x^{2m-2k-1}z^{2k}(H_{2k}-\log z)\varphi(2m-2k)$$
$$+\sum_{k=0}^{m-2}\frac{(-1)^k\zeta(2m-2k-1)}{(2k+1)!}\sum_{j=0}^{k}\binom{2k+1}{2j+1}z^{2k-2j}x^{2j+1}\varphi(2j+2),\tag{101}$$

As for the remainder of the first series in Eq. (91), by Eq. (97), we find

$$\lim_{\alpha\to 2m}\sum_{k=m}^{\infty}\frac{(\alpha-1)\cdots(\alpha-2m)\dfrac{\pi}{2}\cdots(\alpha-2k-1)z^{\alpha-1}\left(\dfrac{x}{z}\right)^{2k+1}}{(-1)^m\Gamma(\alpha)(2k+1)!\sin(\alpha-2m)\dfrac{\pi}{2}}(H_{2k+1}-H_k-\log 2)$$
$$=\frac{(-1)^{m-1}z^{2m-1}}{(2m)!}\sum_{k=m}^{\infty}\frac{\left(\frac{x}{z}\right)^{2k+1}}{\binom{2k+1}{2m}}(H_{2k+1}-H_k-\log 2)$$
$$=\frac{(-1)^{m-1}z^{2m-1}}{(2m)!}\sum_{k=m}^{\infty}\frac{\left(\frac{x}{z}\right)^{2k+1}}{\binom{2k+1}{2m}}(H_{2k+1}-H_k)$$
$$+\frac{(-1)^m x^{2m+1}\log 2}{z^2(2m+1)!}\,{}_3F_2\left(1,1,\frac{3}{2};m+1,m+\frac{3}{2};\frac{x^2}{z^2}\right).\tag{102}$$

If $\Phi_{2m}^{(2)}(x,z)$ denotes the last line expression, we obtain

$$I_{2m}^{S,\cos}=\Phi_{2m}^{(1)}(x,z)+\Phi_{2m}^{(2)}(x,z)+\frac{(-1)^{m-1}z^{2m-1}}{(2m)!}\sum_{k=m}^{\infty}\frac{\left(\frac{x}{z}\right)^{2k+1}}{\binom{2k+1}{2m}}(H_{2k+1}-H_k)$$
$$+\sum_{k=m}^{\infty}(-1)^k\zeta(2m-2k-1)\sum_{j=0}^{k}\frac{z^{2k-2j}x^{2j+1}}{(2k-2j)!(2j+1)!}(H_{2j+1}-H_j-\log 2).\tag{103}$$

Introducing a new index $n=k-m+1$ in the second series and relying on the relation $(2\pi)^{2n}\zeta(1-2n)=(-1)^n2(2n-1)!\zeta(2n)$, Eq. (103) becomes

$$I_{2m}^{S,\cos} = \Phi_{2m}^{(1)}(x,z) + \Phi_{2m}^{(2)}(x,z) + \frac{(-1)^{m-1}z^{2m-1}}{(2m)!}\sum_{k=m}^{\infty}\frac{\left(\frac{x}{z}\right)^{2k+1}}{\binom{2k+1}{2m}}(H_{2k+1} - H_k),$$

$$+2(-1)^{m-1}\sum_{n=1}^{\infty}\left(\frac{z}{2\pi}\right)^{2n}(2n-1)!\zeta(2n)\sum_{j=0}^{n+m-1}\frac{z^{2m-1}\left(\frac{x}{z}\right)^{2j+1}(H_{2j+1} - H_j - \log 2)}{(2n+2m-2-2j)!(2j+1)!}. \tag{104}$$

We split the second series of (104) in two, and the part containing $-\log 2$, we bring to the closed form

$$\begin{aligned}\Phi_{2m}^{(3)}(x,z) = {} & \frac{(-1)^m \log 2}{(2m-1)!z^2}((H_{2m-1} - \log 2\pi)\sum_{k=0}^{m-1}\binom{2m-1}{2j+1}z^{2m-2j-2}x^{2j+1} \\ & + \sum_{k=0}^{m-2}(-1)^{m+k-1}\frac{(2m-1)!}{(2k+1)!}\zeta(2m-2k-1)\sum_{j=0}^{k}\binom{2k+1}{2j+1}z^{2k-2j}x^{2j+1} \\ & -\frac{1}{2}(2\pi)^{2m-1}\left(\zeta'\left(1-2m, 1+\frac{x-z}{2\pi}\right) - \zeta'(1-2m, 1-\frac{x-z}{2\pi}\right) \\ & +\zeta'\left(1-2m, 1+\frac{x+z}{2\pi}\right) - \zeta'(1-2m, 1-\frac{x+z}{2\pi})\Big),\end{aligned} \tag{105}$$

where $\zeta(s,a)$ is Hurwitz's zeta function. Thus, by adding (105), we can write $\Phi_{2m}(x,z) = \Phi_{2m}^{(1)}(x,z) + \Phi_{2m}^{(2)}(x,z) + \Phi_{2m}^{(3)}(x,z)$ and arrive at the following summation formula:

$$\begin{aligned}I_{2m}^{S,\cos} = {} & \Phi_{2m}(x,z) + (-1)^{m-1}z^{2m-1}\left(\sum_{k=m}^{\infty}\frac{\left(\frac{x}{z}\right)^{2k+1}(2k+1-2m)!(H_{2k+1} - H_k)}{(2k+1)!}\right. \\ & \left. +2\sum_{n=1}^{\infty}\left(\frac{z}{2\pi}\right)^{2n}(2n-1)!\zeta(2n)\sum_{j=0}^{n+m-1}\frac{\left(\frac{x}{z}\right)^{2j+1}(H_{2j+1} - H_j)}{(2n+2m-2-2j)!(2j+1)!}\right).\end{aligned} \tag{106}$$

The sum of the series Eq. (82) is not a closed-form expression; nevertheless, this is another way of representing the series (106).

4.3 A more general series

We consider here a series of the form

$$I_{\alpha,\omega}^{T,g} = \sum_{n=1}^{\infty}\frac{(s)^{n-1}T((an-b)x)}{(an-b)^{\alpha}\left((an-b)^2 - \omega^2\right)}g((an-b)z), \quad \alpha \in \mathbb{N}, \tag{107}$$

where $a = \begin{Bmatrix}1\\2\end{Bmatrix}$ $b = \begin{Bmatrix}0\\1\end{Bmatrix}$, $s = 1$ or -1, $\omega \in \mathbb{R}$, $|\omega| \neq an - b$, and $T(an-b)x)$ is the Fourier sine or cosine integral Eq. (9).

We shall consider now the series Eq. (107) for $\alpha = 2m + p - 1, m \in \mathbb{N}$, where p equals 0 or 1, that is,

$$I^{T,g}_{2m+p-1,\omega} = \sum_{n=1}^{\infty} \frac{(s)^{n-1} T((an-b)x) g((an-b)z)}{(an-b)^{2m+p-1}\left((an-b)^2 - \omega^2\right)}, \tag{108}$$

where $\omega| \neq an - b, m \in \mathbb{N}_0$.

Theorem 1.6 The summation formula of the series Eq. (108) is

$$I^{T,g}_{2m+p-1,\omega} = \frac{sp(1-b)}{2\omega^{2m+2}} T(0) - \frac{s\pi \sin^b \frac{\pi\omega}{2}}{2\omega^{2m+p} \sin \pi\omega} g_\omega(z) T(\omega xy) - \sum_{k=1}^{m} \frac{I^{T,g}_{2k+p-1}}{\omega^{2m-2k+2}}, \tag{109}$$

where $I^{T,g}_{p-1,\omega}$ is the sum of Eq. (108) for $m = 0$ and $I^{T,g}_{2k+p-1}$ is the sum of Eq. (93) for $m = k$ and $r = 1 - p$.

Proof. Relying on the decomposition similar to Eq. (23)

$$\frac{1}{\omega^{2m} n^{p-1}(n^2 - \omega^2)} - \frac{1}{n^{2m+p-1}(n^2 - \omega^2)} = \sum_{k=1}^{m} \frac{1}{\omega^{2m-2k+2} n^{2k+p-1}}, \tag{110}$$

where p equals 0 or 1. We proclaim n a summation index running over positive integers and, making use of Eq. (110), obtain the relation

$$\sum_{n=1}^{\infty} \frac{1}{n^{2m+p-1}(n^2 - \omega^2)} = \frac{1}{\omega^{2m}} \sum_{n=1}^{\infty} \frac{1}{n^{p-1}(n^2 - \omega^2)} - \sum_{k=1}^{m} \frac{1}{\omega^{2m-2k+2}} \sum_{n=1}^{\infty} \frac{1}{n^{2k+p-1}}. \tag{111}$$

Multiplying Eq. (110) by $T(nx)g(nz)$, then taking $an - b$ instead of n and implementing (111), we represent Eq. (108) as a difference of two series

$$\begin{aligned} \sum_{n=1}^{\infty} \frac{T((an-b)x) g((an-b)z)}{(an-b)^{2m+p-1}\left((an-b)^2 - \omega^2\right)} &= \frac{1}{\omega^{2m}} \sum_{n=1}^{\infty} \frac{T((an-b)x) g((an-b)z)}{(an-b)^{p-1}\left((an-b)^2 - \omega^2\right)} \\ &- \sum_{k=1}^{m} \frac{1}{\omega^{2m-2k+2}} \sum_{n=1}^{\infty} \frac{T((an-b)x) g((an-b)z)}{(an-b)^{2k+p-1}}. \end{aligned} \tag{112}$$

If $I^{T,f}_{p-1,\omega}$ and $I^{T,g}_{2k+p-1}$ stand for two right-hand side series of Eq. (112), respectively, we rewrite the series Eq. (108) as follows:

$$I^{T,g}_{2m+p-1,\omega} = \frac{I^{T,f}_{p-1,\omega}}{\omega^{2m}} - \sum_{k=1}^{m} \frac{I^{T,g}_{2k+p-1}}{\omega^{2m-2k+2}}, \tag{113}$$

That means, we first consider the series Eq. (108) for $m = 0$, and then apply the formula Eq. (93), putting there $2k + p - 1$ in place of $2m - r$.

So, by setting Fourier sine or cosine integral Eq. (9) in the first series, after interchanging summation and integration, we have

$$I^{T,f}_{p-1,\omega} = \int_0^{\infty} \left(\sum_{n=1}^{\infty} \frac{(s)^{n-1} f((an-b)xy) g((an-b)z)}{(an-b)^{p-1}\left((an-b)^2 - \omega^2\right)} \right) \phi(y)\, \mathrm{d}y. \tag{114}$$

We replace the above series with the formula [28] for the trigonometric series, where we take xy instead of x and z instead of y. Note that convergence regions for this series are those on p. 16. Thus, taking account of Eq. (9), we have

$$\begin{aligned} I_{p-1,\omega}^{T f} &= \int_0^{\infty}\left(f(0)\frac{sp(1-b)}{2\omega^2} - \frac{s\pi \sin^b \frac{\pi\omega}{2}}{2\omega^p \sin \pi\omega} f(\omega x y) g_{\omega}(z)\right)\phi(y)\,\mathrm{d}y \\ &= \frac{sp(1-b)}{2\omega^2} T(0) - \frac{s\pi \sin^b \frac{\pi\omega}{2}}{2\omega^p \sin \pi\omega} g_{\omega}(z) T(\omega x y), \end{aligned} \tag{115}$$

where $g_{\omega}(z) = g\left(\omega z - \frac{\pi}{2a}(s+1)(b+\omega)\right)$, and Eq. (115) presents the first two terms on the right-hand side of the formula Eq. (109).

Now, we use the closed-form formula Eq. (93), replacing r with $1-p$ and m with k, but taking the running index j instead. Thus, the sum of the second right-hand side series in Eq. (112) is $I_{2k+p-1}^{T,g}$. Thereby, we obtain Eq. (109). □

Example 4.2 Choosing $T(x) = C(x)$, $g(y) = y$, for $0 < y < 1$, $g(y) = 0$ for $y > 1$, $\omega = \frac{1}{2}, g = \sin, m = 2, p = 0, s = -1, a = 1, b = 0$ in Eq. (109), there has to be $c = 0, F = \eta$, so we obtain

$$\begin{aligned} I_{3,\frac{1}{2}}^{C,\sin} &= \sum_{n=1}^{\infty} \frac{(-1)^{n-1}}{n^3\left(n^2 - \left(\frac{1}{2}\right)^2\right)} \sin nz \int_0^1 y \cos nxy\,\mathrm{d}y \\ &= \frac{16\pi}{x}\sin\frac{x}{2}\sin\frac{z}{2} + \frac{32\pi}{x^2}\cos\frac{x}{2}\sin\frac{z}{2} - \frac{32\pi}{x^2}\sin\frac{z}{2} - 4z - \frac{\pi^2 z}{6} + \frac{z^3}{6} + \frac{x^2 z}{4}. \end{aligned} \tag{116}$$

This way, by Eq. (116), we obtain a finite sum of the Schlömilch series over a product of the sine series and the Fourier cosine transform.

5. Product of the Hankel integral and a trigonometric function

Here, we construct a new series by multiplying the Hankel integral transform and a trigonometric function, which means we consider the series [34].

$$I_{\alpha}^{D_{\nu} g} = \sum_{n=1}^{\infty} \frac{(s)^{n-1} D_{\nu}((an-b)x)}{(an-b)^{\alpha}} g((an-b)z), \quad \alpha \in \mathbb{R}^{+}, \tag{117}$$

where $a = \begin{Bmatrix} 1 \\ 2 \end{Bmatrix}$ $b = \begin{Bmatrix} 0 \\ 1 \end{Bmatrix}$, $s = 1$ or -1, $g = \sin$ or $g = \cos$, and, $D_{\nu}(x)$ denotes the Hankel-type integrals $B_{\nu}(x)$ or $S_{\nu}(x)$ defined by Eq. (22).

Theorem 1.7 For $\sigma = \alpha - \nu$, the summation formula for the series Eq. (117)

$$\begin{aligned} I_{\alpha}^{D_{\nu} g} &= \frac{\tau\left(\frac{x}{2}\right)^{\nu}}{\sqrt{\pi}}\left(\frac{c\pi}{2\Gamma(\sigma) h\left(\frac{\pi\sigma}{2}\right)} \sum_{k=0}^{\infty}\binom{\sigma-1}{2k+d} \frac{z^{\sigma-1-2k-d} x^{2k+d}\,\Gamma\left(k+\frac{d+1}{2}\right)}{\Gamma\left(\nu+k+1+\frac{d}{2}\right)}\Phi_1\right. \\ &\left. + \sum_{k=0}^{\infty}\frac{(-1)^k F(\sigma-2k-\delta)}{(2k+\delta)!}\sum_{j=0}^{k-v}\binom{2k+\delta}{2j+d}\frac{z^{2k+\delta-2j-d} x^{2j+d}\,\Gamma\left(j+\frac{d+1}{2}\right)}{\Gamma\left(\nu+j+1+\frac{d}{2}\right)}\Phi_2\right), \end{aligned} \tag{118}$$

where $\varphi_\nu = \begin{Bmatrix} J_\nu \\ \mathbf{H}_\nu \end{Bmatrix} f = \begin{Bmatrix} \cos \\ \sin \end{Bmatrix} d = \begin{Bmatrix} 0 \\ 1 \end{Bmatrix}, \begin{Bmatrix} f = g \\ f \neq g \end{Bmatrix} \delta = \begin{Bmatrix} 0 \\ 1 \end{Bmatrix} h = \begin{Bmatrix} \cos \\ \sin \end{Bmatrix},$ $\tau = -1$, and $v = 1$ only if $f = g = \sin$; otherwise, $\tau = 1$ and $v = 0$. The parameters a, b, s, c, F and convergence regions K_1, K_2, K_3, K_4 are on p. 16, and $\Phi_1 = \varphi(\nu + 2k + d + 2)$ and $\Phi_2 = \varphi(\nu + 2j + d + 2)$ are the Mellin transforms.

Proof. To find the sum, we do not first calculate the integral $D(x)$, but place Eq. (22) in Eq. (117) instead. Afterward, we rely on interchanging summation and integration, so the procedure of obtaining a summation formula for the series Eq. (117) reduces to finding a sum of the series over a product of the Bessel or Struve function and a trigonometric function, sine or cosine, that is,

$$\begin{aligned} I_\alpha^{D_\nu g} &= \sum_{n=1}^{\infty} \frac{(s)^{n-1} g((an-b)z)}{(an-b)^\alpha} \int_0^\infty y\phi(y)\varphi_\nu((an-b)xy)\,dy \\ &= \int_0^\infty \left(\sum_{n=1}^{\infty} \frac{(s)^{n-1} g((an-b)z)}{(an-b)^\alpha} \varphi_\nu((an-b)xy) \right) y\phi(y)\,dy, \end{aligned} \tag{119}$$

with φ_ν as a Bessel J_ν or Struve function $\mathbf{H}_\nu$. By confirming uniform convergence of the right-hand series in Eq. (119) with respect to $y \in (0, \infty)$, we shall justify Eq. (119). For this purpose, we use an integral representation of the Bessel or Struve function [2, p. 224, entry 10.9.4] where $\nu > -\frac{1}{2}$, $\varphi_\nu = \begin{Bmatrix} J_\nu \\ \mathbf{H}_\nu \end{Bmatrix} f = \begin{Bmatrix} \cos \\ \sin \end{Bmatrix}$. Further, we place it instead of φ_ν in the right-hand side series of Eq. (119), then we may interchange integration and summation

$$\begin{aligned} S_\alpha^{\varphi_\nu g} &= \frac{2\left(\frac{xy}{2}\right)^\nu}{\sqrt{\pi}\Gamma\left(\nu + \frac{1}{2}\right)} \sum_{n=1}^{\infty} \frac{(s)^{n-1} g((an-b)z)}{(an-b)^{\alpha-\nu}} \int_0^1 \left(1 - t^2\right)^{\nu - \frac{1}{2}} f((an-b)xyt)\,dt \\ &= \frac{2\left(\frac{xy}{2}\right)^\nu}{\sqrt{\pi}\Gamma\left(\nu + \frac{1}{2}\right)} \int_0^1 \left(1 - t^2\right)^{\nu - \frac{1}{2}} \sum_{n=1}^{\infty} \frac{(s)^{n-1} f((an-b)xyt) g((an-b)z)}{(an-b)^{\alpha-\nu}}\,dt, \end{aligned} \tag{120}$$

because the last series uniformly converges with respect to $y \in (0, \infty)$ (see the right-hand series in Eq. (84), p. 16). So, the convergence regions of the series Eq. (117) are discussed on p. 16, where y ought to be replaced by z.

We are referring now to [23], p. 356, Eq. (4.1), which is the sum of the right-hand series in Eq. (119), and change there y to z and x to xy. For $\alpha - \nu = \sigma$, we have

$$\begin{aligned} S_\alpha^{\varphi_\nu g} &= \frac{c\tau\left(\frac{x}{2}\right)^\nu \sqrt{\pi}}{2\Gamma(\sigma) h\left(\frac{\sigma\pi}{2}\right)} \sum_{k=0}^{\infty} \begin{pmatrix} \sigma - 1 \\ 2k + d \end{pmatrix} \frac{z^{\sigma-1-2k-d} x^{2k+d} \Gamma\left(k + \frac{d+1}{2}\right)}{\Gamma\left(\nu + k + 1 + \frac{d}{2}\right)} y^{\nu+2k+d} \\ &+ \sum_{k=0}^{\infty} \frac{\tau(-1)^k F(\sigma - 2k - \delta)}{2^\nu \sqrt{\pi}(2k+\delta)!} \sum_{j=0}^{k-v} \begin{pmatrix} 2k + \delta \\ 2j + d \end{pmatrix} \frac{z^{2k+\delta-2j-d} (xy)^{\nu+2j+d} \Gamma\left(j + \frac{d+1}{2}\right)}{\Gamma\left(\nu + j + 1 + \frac{d}{2}\right)}. \end{aligned} \tag{121}$$

So, by replacing the right-hand side of Eq. (119) with $S_\alpha^{\varphi_\nu g}$ from (121), after some rearrangements, we obtain Eq. (118). □

5.1 Closed-form cases

The formula Eq. (118) is brought in closed form if $\sigma = \alpha - \nu \in \mathbb{N}_0$ and the function F vanishes. So, for $\sigma = 2m - p, m \in \mathbb{N}_0$, with $p = 1$ or $p = 0$, it becomes

$$\frac{\left(\frac{x}{2}\right)^{\nu}\tau}{\sqrt{\pi}}\left(\frac{c\pi}{2\Gamma(\sigma)h\left(\frac{\sigma\pi}{2}\right)}\sum_{k=0}^{m-1}\binom{\sigma-1}{2k+d}\frac{z^{\sigma-1-2k-d}x^{2k+d}\Gamma\left(k+\frac{d+1}{2}\right)}{\Gamma\left(k+\nu+1+\frac{d}{2}\right)}\Phi_1\right.$$
$$\left.+\sum_{k=0}^{m-1}\frac{(-1)^k F(\sigma-2k-\delta)}{(2k+\delta)!}\sum_{j=0}^{k-\nu}\binom{2k+\delta}{2j+d}\frac{z^{2k+\delta-2j-d}x^{2j+d}\Gamma\left(j+\frac{d+1}{2}\right)}{\Gamma\left(j+\nu+1+\frac{d}{2}\right)}\Phi_2\right), \tag{122}$$

with the Mellin transforms $\Phi_1 = \varphi(\nu + 2k + d + 2), \Phi_2 = \varphi(\nu + 2j + d + 2)$. All the parameters are the same as for the formula Eq. (118).

Example 5.1 By choosing $D_\nu = B_\nu$ and $g = \sin$, there has to be $\varphi_\nu = J_\nu$ and $f = \cos$, which implies $d = 0$ and $\delta = 1$, $h = \sin$ because $f \neq g$. For $a = 2, b = 1, s = -1$, it is necessary to be $c = 0$, and $F = \beta$. Further, let $\sigma = 2$ and $\phi(y) = y^\mu$ for $0 < y < 1$ and $\phi(y) = 0$ for $1 < y < \infty$. From Eq. (122), we obtain

$$I_{\nu+2}^{B,\sin} = \sum_{n=1}^{\infty}\frac{(-1)^{n-1}B_\nu((2n-1)x)\sin(2n-1)z}{(2n-1)^{\nu+2}} = \frac{z\left(\frac{x}{2}\right)^{\nu}\pi}{4(2+\nu+\mu)\Gamma(\nu+1)}, \tag{123}$$

with $(x, z) \in K_4$. On the other hand, from [6, p. 22, (8)], we derive

$$B_\nu(x) = \int_0^1 y^{\mu+1}J_\nu(xy)\,dy$$
$$= (\nu+\mu)\frac{J_\nu(x)}{x^{\mu+1}}S_{\mu,\nu-1}(x) - \frac{J_{\nu-1}(x)}{x^{\mu+1}}S_{\mu+1,\nu}(x) + \frac{2^{\mu+1}\Gamma\left(1+\frac{\nu+\mu}{2}\right)}{x^{\mu+2}\Gamma\left(\frac{\nu-\mu}{2}\right)}, \tag{124}$$

where $S_{\mu,\nu}$ is the Lommel function (see [27, entry 11.9.5]) and $\mu + \nu > -2$. Taking Eq. (124) to Eq. (123), we obtain

$$\sum_{n=1}^{\infty}\frac{(-1)^{n-1}\sin(2n-1)z}{(2n-1)^{\mu+\nu+3}}((\nu+\mu)J_\nu((2n-1)x)S_{\mu,\nu-1}(2n-1)x)$$
$$-J_{\nu-1}((2n-1)x)S_{\mu+1,\nu}((2n-1)x)) = \frac{x^{\nu+\mu+1}z\pi}{2^{\nu+2}(2+\nu+\mu)\Gamma(\nu+1)}$$
$$-\frac{2^{\mu+1}\Gamma\left(1+\frac{\nu+\mu}{2}\right)}{x\Gamma\left(\frac{\nu-\mu}{2}\right)}\sum_{n=1}^{\infty}\frac{(-1)^{n-1}\sin(2n-1)z}{(2n-1)^{\mu+\nu+4}}. \tag{125}$$

To determine the sum of the trigonometric series, we refer to **Table 1** for $s = -1, a = 2, b = 1, f = \sin, \alpha = \mu + \nu + 4$. So, we find the formula

$$\sum_{n=1}^{\infty}\frac{(-1)^{n-1}\sin(2n-1)z}{(2n-1)^{\mu+\nu+4}}=\sum_{k=0}^{\infty}\frac{(-1)^{k}\beta(\mu+\nu-2k+3)}{(2k+1)!}z^{2k+1}. \tag{126}$$

Thus, we finally obtain the sum of Eq. (125), which is brought in closed form if $\mu+\nu$ is an even integer greater than -4 because β vanishes at negative odd integers. So, in this case, the sum of Eq. (125) is

$$\frac{x^{\nu+\mu+1}z\pi}{2^{\nu+2}(2+\nu+\mu)\Gamma(\nu+1)}-\frac{2^{\mu+1}\Gamma\left(1+\frac{\nu+\mu}{2}\right)}{x\Gamma\left(\frac{\nu-\mu}{2}\right)}\sum_{n=1}^{\frac{\mu+\nu}{2}+1}\frac{(-1)^{k}\beta(\mu+\nu-2k+3)}{(2k+1)!}z^{2k+1}. \tag{127}$$

Example 5.2 Let $a=1$, $b=0$, and $s=-1$ in Eq. (117), implying $c=0$ and $F=\eta$. For $D_\nu=B_\nu$, there must be $f=\cos$ and $d=0$. Further, if we take $g=\cos$ then $\delta=0$ and $h=\cos$ because $f=g$. We choose $\nu=\frac{1}{2}$, $\phi(y)=y^{-1}(1-y^2)^{-1/2}$, and $m=2$. So, applying Eq. (122), we find

$$\begin{aligned}&\sum_{n=1}^{\infty}(-1)^{n-1}\frac{\cos nz}{n^{9/2}}\int_0^1\frac{J_{1/2}(nxy)}{\sqrt{1-y^2}}\,\mathrm{d}y\\&=\frac{\pi\sqrt{x}}{2\Gamma^2\left(\frac{1}{4}\right)}\left(\frac{7\pi^4}{90}-\frac{\pi^2z^2}{3}+\frac{z^4}{6}-\frac{\pi^2x^2}{15}+\frac{x^2z^2}{5}+\frac{7x^4}{450}\right),\end{aligned} \tag{128}$$

where $|x|<\pi$ and $|x|-\pi<z<\pi-|x|$, that is, the K_2 region, p. 16.

We have emphasized that it is not necessary to calculate the integrals (22) to find the sum of the series Eq. (117). Besides, there are good chances one might not do this elementarily. However, based on [4, entry 4.33, p. 38], we find

$$\int_0^1\frac{J_\nu(nxy)}{\sqrt{1-y^2}}\,\mathrm{d}y=\frac{\pi}{2}J^2_{\nu/2}\left(\frac{nx}{2}\right),\quad \nu>-1. \tag{129}$$

Putting Eq. (129) with $\nu=\frac{1}{2}$ in Eq. (128) gives rise to the summation formula

$$\sum_{n=1}^{\infty}\frac{(-1)^{n-1}J^2_{1/4}\left(\frac{nx}{2}\right)\cos nz}{n^{9/2}}=\frac{\sqrt{x}}{\Gamma^2\left(\frac{1}{4}\right)}\left(\frac{7\pi^4}{90}-\frac{\pi^2z^2}{3}+\frac{z^4}{6}-\frac{\pi^2x^2}{15}+\frac{x^2z^2}{5}+\frac{7x^4}{450}\right). \tag{130}$$

This way, we obtain by Eq. (130) a closed form of a series over square Bessel functions.

5.2 Limiting value cases

On the other hand, we cannot immediately set $h=\sin$ and $\alpha-\nu=2m$ or $h=\cos$ and $\alpha-\nu=2m-1$, and $m\in\mathbb{N}$, in the formula Eq. (118), because the first term of Eq. (118) has zero as a divisor, so we have to deal with a limiting value and replace first $\alpha-\nu$ with σ and then take a limit $\sigma\to 2m$ or $\sigma\to 2m-1$.

Example 5.3 We choose, $a=2, b=1, s=1$, so $c=\frac{1}{2}$, $F=\lambda$, and for $\varphi_\nu=\mathbf{H}_\nu$, we set $f=\sin$, and after $g=\cos$, we have $\delta=1, d=1, h=\sin$. Finally, by taking $\phi(y)=y^{-\nu}$ for $0<y<1$ and $\phi(y)=0$ for $y>1$, we find

$$\sum_{n=1}^{\infty}\frac{S_\nu((2n-1)x)}{(2n-1)^\alpha}\cos(2n-1)z=\frac{\left(\frac{x}{2}\right)^\nu\sqrt{\pi}}{4\Gamma(\sigma)\sin\frac{\pi\sigma}{2}}\sum_{k=0}^{\infty}\binom{\sigma-1}{2k+1}\frac{z^{\sigma-2k-2}x^{2k+1}k!}{(2k+3)\Gamma\left(k+\nu+\frac{3}{2}\right)}$$
$$+\frac{\left(\frac{x}{2}\right)^\nu}{\sqrt{\pi}}\sum_{k=0}^{\infty}\frac{(-1)^k\lambda(\sigma-2k-1)}{(2k+1)!}\sum_{j=0}^{k}\binom{2k+1}{2j+1}\frac{z^{2k-2j}x^{2j+1}j!}{(2j+3)\Gamma\left(j+\nu+\frac{3}{2}\right)},\tag{131}$$

To determine a limiting value for the series Eq. (131), we are acting so that we split both series in two, up to $k=m-1$ and from $k=m$. Thus, we first deal with the limit

$$\lim_{\sigma\to 2m}\left(\frac{\left(\frac{x}{2}\right)^\nu\sqrt{\pi}}{4\Gamma(\sigma)\sin\frac{\pi\sigma}{2}}\sum_{k=0}^{m-1}\binom{\sigma-1}{2k+1}\frac{z^{\sigma-2k-2}x^{2k+1}k!}{(2k+3)\Gamma\left(k+\nu+\frac{3}{2}\right)}\right.$$
$$\left.+\frac{\left(\frac{x}{2}\right)^\nu}{\sqrt{\pi}}\sum_{k=0}^{m-1}\frac{(-1)^k\lambda(\sigma-2k-1)}{(2k+1)!}\sum_{j=0}^{k}\binom{2k+1}{2j+1}\frac{z^{2k-2j}x^{2j+1}j!}{(2j+3)\Gamma\left(j+\nu+\frac{3}{2}\right)}\right),\tag{132}$$

and by bringing the corresponding fractions to the same denominator, we obtain a value, which we designate with $L_{2m}(x,z)$

$$L_{2m}(x,z)=\frac{\left(\frac{x}{2}\right)^{\nu+1}}{\sqrt{\pi}}\left(\sum_{k=0}^{m-1}\frac{\left(H_{2m-2k-2}-\log\frac{z}{2}\right)}{(2k+3)\Gamma\left(\nu+k+\frac{3}{2}\right)}\frac{(-1)^{m-1}z^{2m-2k-2}x^{2k}k!}{(2m-2k-2)!(2k+1)!}\right.$$
$$\left.+2\sum_{k=0}^{m-2}\frac{(-1)^k\lambda(2m-2k-1)}{(2k+1)!}\sum_{j=0}^{k}\binom{2k+1}{2j+1}\frac{z^{2k-2j}x^{2j}j!}{(2j+3)\Gamma\left(j+\nu+\frac{3}{2}\right)}\right),\tag{133}$$

with H_n as the nth harmonic number.

Further, we shall find a limiting value for the rest of the first series

$$\lim_{\sigma\to 2m}\frac{\left(\frac{x}{2}\right)^{\nu+1}\sqrt{\pi}}{4\sin\frac{\sigma\pi}{2}\Gamma(\sigma)}\sum_{k=m}^{\infty}\binom{\sigma-1}{2k+1}\frac{z^{\sigma-2k-2}x^{2k}k!}{\Gamma\left(\nu+k+\frac{3}{2}\right)}$$
$$=\frac{\left(\frac{x}{2}\right)^{\nu+1}(-1)^{m-1}}{\Gamma\left(\nu+\frac{3}{2}\right)\sqrt{\pi}}\lim_{\sigma\to 2m}\frac{z^{\sigma-2}}{\Gamma(\sigma-1)}\sum_{n=m}^{\infty}\frac{\left(1-\frac{\sigma}{2}\right)\cdots\left(m-\frac{\sigma}{2}\right)\pi\cdots\left(n-\frac{\sigma}{2}\right)\left(\frac{3-\sigma}{2}\right)_n x^{2n}}{\sin\left(m-\frac{\sigma}{2}\right)\pi\left(\frac{3}{2}\right)_n\left(\nu+\frac{3}{2}\right)_n}\frac{x^{2n}}{z^{2n}}.\tag{134}$$

Let $R(x,z)$ denote the obtained limiting value in Eq. (134), then we have

$$R_{2m}(x,z)=\frac{\left(\frac{x}{2}\right)^{\nu+1}z^{2m-2}}{\Gamma\left(\nu+\frac{3}{2}\right)\sqrt{\pi}}\frac{(m-1)!}{(2m-2)!}\sum_{n=m}^{\infty}\frac{(n-m)!\left(\frac{3}{2}-m\right)_n}{\left(\frac{3}{2}\right)_n\left(\nu+\frac{3}{2}\right)_n}\frac{x^{2n}}{z^{2n}}.\tag{135}$$

The relation Eq. (135) can be rewritten through a hypergeometric function, that is,

$$R_{2m}(x,z) = \frac{(-1)^{m+1} x^{\nu+2m+1} {}_3F_2\left(1, 1, \frac{3}{2}; m+\frac{5}{2}, \nu+m+\frac{3}{2}; \frac{x^2}{z^2}\right)}{2^{\nu+m+1}(2m+3)!!\sqrt{\pi} z^2 \Gamma\left(\nu+m+\frac{3}{2}\right)}. \tag{136}$$

Taking into account Eq. (136) and Eq. (133), the series Eq. (131) is now as follows:

$$\begin{aligned} I_{\nu+2m}^{S_\nu,\cos} &= \sum_{n=1}^{\infty} \frac{S_\nu((2n-1)x)}{(2n-1)^\alpha} \cos(2n-1)z = L_{2m}(x,z) + R_{2m}(x,z) \\ &+ \frac{\left(\frac{x}{2}\right)^\nu}{\sqrt{\pi}} \sum_{k=m}^{\infty} \frac{(-1)^k \lambda(2m-2k-1)}{(2k+1)!} \sum_{j=0}^{k} \binom{2k+1}{2j+1} \frac{z^{2k-2j} x^{2j+1} j!}{(2j+3)\Gamma\left(j+\nu+\frac{3}{2}\right)}. \end{aligned} \tag{137}$$

Again, as with Eq. (106) concerning the series Eq. (82), the formula Eq. (137) offers another representation of the series $I_{\nu+2m}^{S_\nu,\cos}$.

On the other hand, in view of (see [24, Section 11.2, entry(3)]), we find

$$S_\nu(x) = \int_0^1 y^{1-\nu} \mathbf{H}_\nu(xy)\, dy = \frac{\left(\frac{x}{2}\right)^{\nu-1}}{\sqrt{\pi}\Gamma\left(\nu+\frac{1}{2}\right)} - \frac{\mathbf{H}_{\nu-1}(x)}{x}. \tag{138}$$

The identity Eq. (138) implies

$$\begin{aligned} \sum_{n=1}^{\infty} \frac{S_\nu((2n-1)x)}{(2n-1)^{\nu+2m}} \cos(2n-1)z &= \frac{\left(\frac{x}{2}\right)^{\nu-1}}{\sqrt{\pi}\Gamma\left(\nu+\frac{1}{2}\right)} \sum_{n=1}^{\infty} \frac{\cos(2n-1)z}{(2n-1)^{2m+1}} \\ &- \frac{1}{x} \sum_{n=1}^{\infty} \frac{\mathbf{H}_{\nu-1}((2n-1)x)}{(2n-1)^{\nu+2m+1}} \cos(2n-1)z. \end{aligned} \tag{139}$$

Thus, the sum of the Schlömilch series over a product of the Hankel integral and a trigonometric function Eq. (139); one can also determine as a difference between a trigonometric series (see [30]) and a series over a product of Struve's and a trigonometric function (see [23]).

Author details

Slobodan B. Tričković[1]* and Miomir S. Stanković[2]

1 University of Niš, Department of Mathematics, Niš, Serbia

2 Mathematical Institute of the Serbian Academy of Sciences and Arts, Belgrade, Serbia

*Address all correspondence to: sbt@junis.ni.ac.rs

References

[1] Rusev P. Representation of holomorphic functions by Schlömilch's series. Fractional Calculus and Applied Analysis. 2013;**16**(2):431-435

[2] Abramowitz M, Stegun A. Handbook of Mathematical Functions. New York: Cambridge Universiy Press; 2010

[3] Rayleigh JWS. On a physical interpretation of Schlömilch's theorem in Bessel's functions. The London, Edinburgh, and Dublin Philosophical Magazine and Journal of Science. 1911;**21** (124):567-571

[4] Oberhettinger F. Tables of Bessel Transform. New York Heidelberg Berlin: Springer-Verlag; 1972

[5] Magnus W, Oberhettinger F, Tricomi FG. Tables of Integral Transform. Vol. I. New York Toronto London: McGraw-Hill Book Company, Inc.; 1954

[6] Magnus W, Oberhettinger F, Tricomi FG. Tables of Integral Transform. Vol. II. New York Toronto London: McGraw-Hill Book Company, Inc.; 1954

[7] Mali P, Gombar S, Radošević S, Rutonjski M, Pantić M, Pavkov-Hrvojević M. Summation formulas generated by Hilbert space eigenproblem. Journal of Mathematical Physics. 2024;**64**:032104

[8] Tričković SB, Vidanović MV, Stanković MS. On the summation of series over a product of Bessel functions. Integral Transforms and Special Functions. 2006;**17**:749-758

[9] Miller AR. On certain Schlömilch-type series. Journal of Computational and Applied Mathematics. 1997;**80**:83-95

[10] Mali P, Rutonjski M, Radošević S, Pantić M, Pavkov-Hrvojević M. Using quantum mechanics for calculation of different infinite sums. European Journal of Physics. 2022;**43**:035405

[11] Fikioris G. Integral evaluation using the Mellin transform and generalized Hypergeometric functions: Tutorial and applications to antenna problems. IEEE Transactions on Antennas and Propagation. 2006;**54**(12):3895-3907

[12] Schiff JL. The Laplace Transform: The Theory and Applications. New York: Springer-Verlag; 1999

[13] Debnath L, Bhatta D. Integral Transforms and their Applications. 3rd ed. Boca Raton-London-New York: CRS Press Taylor & Francis Group; 2015

[14] Arfken GB, Weber HJ, Harris FE. Mathematical Methods for Physicists. Amsterdam: Elsevier Inc; 2013

[15] Graf U. Introduction to Hyperfunctions and their Integral Transforms. Birkhäuser/Springer Basel AG; 2010

[16] Paris RB, Kaminsky D. Asymptotics and Mellin-Barnes Integrals. Cambridge: Cambridge University Press; 2001

[17] Andrews GE, Askey R, Roy R. Special Functions, Encyclopedia of Mathematics and its Applications 71. Cambridge: Cambridge University Press; 1999

[18] Piessens R. In: Poularikas AD, editor. The Hankel Transform in the Transforms and Applications Handbook. Second ed. Boca Raton: CRC Press LLC; 2000

[19] de Branges L. Hilbert Spaces of Entire Functions. London: Prentice-Hall; 1968

[20] Prudnikov AP, Brychkov YA, Marichev OI. Integrals and Series, Vol. 1, Elementary Functions. Vol. 1. London: Taylor & Francis; 1998

[21] Stanković MS, Vidanović MV, Tričković SB. Closed form expressions for some series over certain trigonometric integrals. Applicable Analysis. 2001;**80**(1–2):53-64

[22] Knopp K. Theory and Application of Infinite Series. New York: Dover; 1990

[23] Tričković SB, Stanković MS, Vidanović MV. On the summation of Schlömilch series. Integral Transforms and Special Functions. 2020;**31**(5): 339-367

[24] Bateman H, Erdélyi A. Tables of Integral Transforms. Vol. II. New York-Toronto-London: McGraw-Hill Book Company, Inc.; 1954

[25] Paris RB. Digital Library of Mathematical Functions, § 11.2 Struve and Related Functions. Available from: https://dlmf.nist.gov/11.2

[26] Jankov-Maširević D. Summations of Schlömilch series containing some Lommel functions of the first kind terms. Integral Transforms and Special Functions. 2016;**27**(2):153-162

[27] Paris RB. Digital Library of Mathematical Functions, § 11.9 Lommel Functions. Available from: https://dlmf.nist.gov/11.9

[28] Stanković MS, Vidanović MV, Tričković SB. On the summation of series involving Bessel or Struve functions. Journal of Mathematical Analysis and Applications. 2000;**247**: 15-26

[29] Stanković MS, Vidanović MV, Tričković SB. Summation of some trigonometric and Schlömilch series. Journal of Computational Analysis and Applications. 2001;**5**(3):313-331

[30] Tričković SB, Stanković MS, Vidanović MV. On the summation of trigonometric series. Integral Transforms and Special Functions. 2008; **19**(6):441-452

[31] Gradshtein IS, Ryzhik IM. In: Jeffrey A, Zwillinger D, editors. Tables of Integrals, Series and Products. Amsterdam: Elsevier, Academic Press; 2007

[32] Tričković SB, Vidanović MV, Stanković MS. Series involving the product of trigonometric integral and trigonometric function. Integral Transforms and Special Functions. 2007; **18**(10):751-763

[33] Oberhettinger F. Tables of Mellin Transforms. Berlin-Heidelberg-NewYork: Springer-Verlag; 1974

[34] Tričković SB, Vidanović MV, Stanković MS. On trigonometric series over integrals involving Bessel or Struve functions. Integral Transforms and Special Functions. 2009;**20**(11):821-834

Chapter 5

Evolutionary and Nature-Inspired Algorithms for Disaster-Resilient Networks

Bidyarani Langpoklakpam and Lithungo K. Murry

Abstract

Disaster management system necessitates efficient and resilient communication networks to ensure effective emergency response and recovery efforts. Disasters pose significant challenges to communication infrastructures, often leading to breakdowns in communication networks and disrupting rescue and relief efforts. In recent years, metaheuristic algorithms have emerged as a promising solution for optimizing various aspects of disaster scenarios. In this paper, we investigate the use and application of metaheuristic algorithms for addressing various optimization problems that arise during emergency operations. The key challenges in disaster management design, including victim localization, routing, coverage, and resource allocation, are discussed. This study also discusses the strengths and limitations of different metaheuristic algorithms for disaster scenarios. Finally, it highlights recently developed models and future research directions in the application of metaheuristic algorithms for disaster area network optimization.

Keywords: disaster management, metaheuristic algorithm, evolutionary algorithm, coverage, resource allocation

1. Introduction

Natural and man-made disasters affect millions of people globally, often resulting in loss of human life and significant damage to infrastructure and property. In recent years, the frequency and severity of such events have highlighted the critical importance of effective disaster management strategies. Most disasters are unforeseen occurrences that cause large numbers of fatalities, property damage, devastation of the environment, and significant effects on society [1]. In the event of such disasters, the ground network infrastructure and road connectivity frequently experience significant disruptions, making it extremely difficult to conduct efficient search and rescue missions [2]. When a disaster occurs, the most critical challenge is to promptly establish initial communication with affected victims who are trapped or inaccessible to rescue teams. Emergency responders can act quickly and effectively when they have efficient emergency management strategies and response protocols, which can help prevent a crisis from escalating [3]. Designing a suitable model for rescuer assignment is crucial to efficiently allocate rescuers to evacuees injured in real-time, considering the dynamic and hazardous environment. However, these tasks are NP-hard because of the dynamic environments and complex relationships between pedestrian movement and information dissemination, which involve minimizing the overall potential cost of the rescue operation [4]. In recent times, researchers have explored metaheuristic algorithms for disaster management, showing significant promise in addressing challenges in disaster applications.

Optimization, a field at the intersection of mathematics and computer science, explores methodologies and techniques for discovering the optimal solution to a given optimization problem. Applying mathematical programming approaches to large-scale problems can be challenging and time-consuming. In response, over the past decades, numerical optimization techniques, particularly metaheuristic algorithms, have gained widespread adoption for addressing real-world problems [5].

Metaheuristic algorithms (MAs) are a subset of heuristic methods, employ iterative processes that explore the entire problem space using various heuristic approaches. In contrast to traditional algorithms, metaheuristic algorithms are usually inspired by theoretical and empirical sources, they frequently mimic biological or natural phenomena [6]. These algorithms are flexible enough to be designed, integrated, or adjusted to fit particular problem domains. By utilizing stochastic techniques, these algorithms methodically navigate intricate search spaces, offering benefits related to resilience and worldwide search abilities [7]. The challenges in disaster management are classified as localization, routing, coverage, and resource allocation. These challenges can be defined as optimization problems, where researchers seek to achieve the best possible results with limited resources. The objective functions can either be optimally maximized or minimized. Various types of objectives for disaster management are illustrated in **Figure 1**.

This study aims to present a thorough analysis of metaheuristic algorithms developed for disaster management system. The paper begins by presenting an overview of MAs and explored the problems involved in emergency and relief operations using MAs. Furthermore, this work provides the strengths and limitations of employing metaheuristic algorithms in disaster management. Additionally, recent optimization

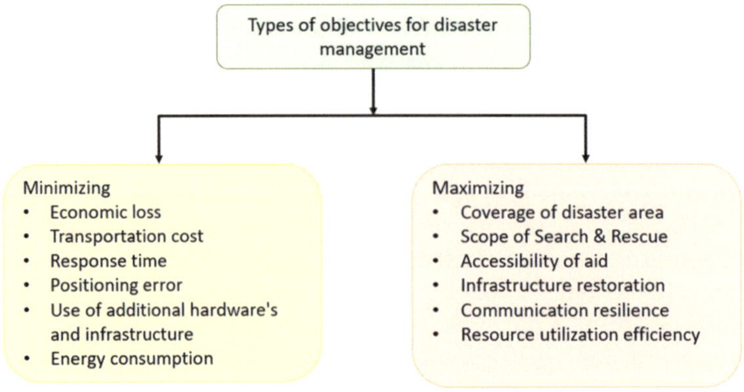

Figure 1.
Various types of objectives for disaster management system.

strategies and hybrid techniques developed using MAs for disaster management are also included in the study. We explore the current status of MAs in disaster management and discuss future directions that may be explored. This paper highlights the significance of MAs for emergency and disaster management by providing insights into the benefits and challenges of MAs.

In the following sections of the paper, we first classify metaheuristic algorithm types into five categories, as outlined in Section 2. Following this, we explore the recent studies in MAs designed to address post-disaster challenges, which are elaborated in Section 3, and hybrid optimization algorithms in Section 4. Section 5 provides the limitations and discussion of the literature review. Section 6 describes the benefits and limitations. Finally, Section 7 discusses the conclusion and future challenges uncovered through this study.

2. Classification of metaheuristic algorithm

Metaheuristics are algorithmic frameworks used across various optimization problems, often requiring minimal modifications to adapt to specific problem instances. In this section, MAs are classified into evolutionary algorithm (EA), physical-based algorithm (PA), nature-inspired algorithm (NA), swarm-based algorithm (SA), and miscellaneous algorithm (MA) as shown in **Figure 2**.

2.1 Evolutionary algorithm

Evolutionary algorithms (EAs) represent a class of modern heuristic search techniques, leveraging the concept of "survival of the fittest" from natural

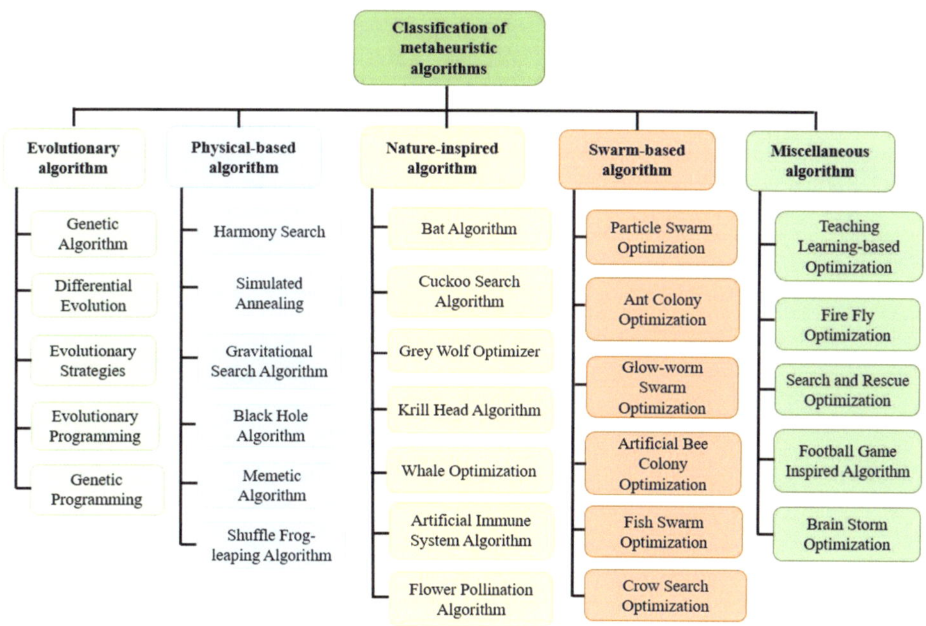

Figure 2.
Classifications of metaheuristic algorithms.

evolution. These algorithms mimic natural processes like genetic inheritance and the Darwinian struggle for survival [8]. Evolutionary-based metaheuristic algorithms, such as genetic algorithms (GA) [9], differential evolution (DE) [10], evolutionary strategies (ES) [11], evolutionary programming (EP) [12], and genetic programming (GP) [13], are inspired by principles of natural evolution and genetics. In disaster area network, EAs can be used to optimize resource allocation strategies and routing protocols for communication networks. Moreover, EAs can handle multi-objective optimization problems effectively, balancing multiple criteria such as minimizing response time, maximizing resource utilization, and minimizing costs.

The EAs optimization process typically involves initializing a population of potential solutions, evaluating each solution's fitness based on an objective function, selecting candidates for reproduction based on their fitness, and creating offspring through recombination or crossover. Offspring then replace some candidates in the current population, maintaining a constant population size. This iterative process continues until a termination criterion is met, such as reaching an optimal solution or running out of computational resources. The steps involved in EAs are shown in **Figure 3**.

EAs maintain a population of potential solutions, inspiring diversity and reducing the likelihood of getting trapped in local optima. It is crucial for disaster relief operations where multiple feasible strategies may need to be evaluated.

2.2 Physical-based algorithms

Physical metaheuristic algorithms are a class of optimization techniques inspired by physical processes and phenomena found in nature. These algorithms simulate physical processes to guide the search for optimal solutions in complex optimization scenarios. The primary key sources of inspiration for this subcategory are gravity, the big bang, black holes, galaxies, and fields [14]. Some popular physical metaheuristic algorithms include harmony search (HS) [15], simulated annealing (SA) [16], gravitational search algorithm (GSA) [17], Black Hole Algorithm (BHA) [18], memetic algorithm (MA) [19], and shuffle frog-leaping algorithm (SFA) [20]. Physical-based optimization can optimize the positioning of communication nodes or facility locations by considering factors such as population distribution, accessibility, terrain

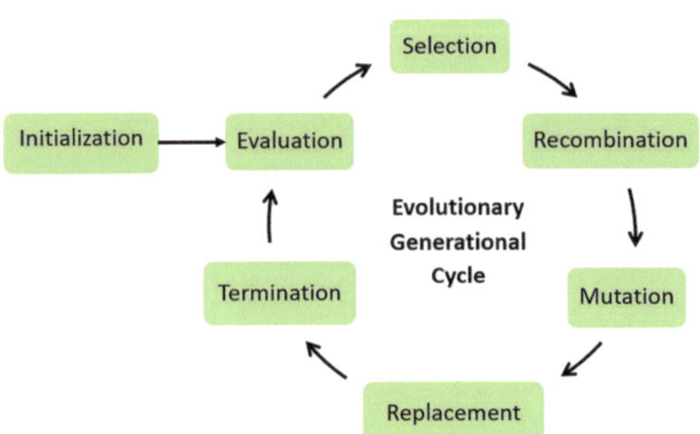

Figure 3.
Steps involve in evolutionary algorithm.

features, and infrastructure availability. Moreover, these techniques can dynamically distribute resources to regions with the most critical requirements by continuously monitoring parameters such as signal strength, traffic volume, and changes in network topology.

2.3 Nature-inspired algorithm

Nature-inspired algorithms are computational approaches that draw inspiration from concepts found in biological systems, ecological systems, and evolutionary mechanisms. These techniques are increasingly applied in disaster management to address challenges such as localization, resource allocation, routing, and communication recovery [21]. Bat Algorithm (BA) [22], Cuckoo Search Algorithm (CSA) [23], Gray Wolf Optimizer (GWO) [24], Flower Pollination Algorithm [25], Whale optimization [26], Krill Head Algorithm (KHA) [27], Artificial Immune System (AIS) [28] are some popular nature-inspired algorithms. Several problems have been effectively tackled by these algorithms in a variety of domains, such as engineering, finance, robotics, optimization, and data mining. These algorithms use the inherent intelligence of nature like the swarming behavior of birds or the foraging behavior of ants to provide effective and efficient solutions for complex problems. Moreover, these techniques can optimize resource allocation, load balancing, reduce the effects of disasters, enhance responder coordination and communication, and quickly deliver vital services to impacted populations.

2.4 Swarm intelligence

The term "swarm intelligence" describes the collective behavior of self-organizing, decentralized systems that are modeled after social insect colonies like termites, ants, and bees. Problems like routing, optimization, and decision-making can be resolved by swarm intelligence systems without the need for explicit coordination or centralized control. Many MAs have been influenced by SI; for example, Particle Swarm Optimization (PSO) [29], Ant Colony Optimization (ACO) [30], Glowworm Swarm Optimization (GSO) [31], Artificial Bee Colony (ABC) [32], Fish Swarm Algorithm (FSA) [33], Crow- Search Algorithm (CSA) [34]. Swarm-based optimization algorithms can facilitate optimal resource allocation in disaster areas, including the deployment of Unmanned Aerial Vehicles (UAVs) for relaying messages and establishing ad hoc networks. By considering factors such as network coverage and energy consumption, these algorithms can help maximize communication range and reliability while minimizing resource usage. These algorithms can optimize search patterns, identify high-priority locations based on survivor likelihood, and adapt to dynamic environments.

The SI algorithms involve several key steps as shown in **Figure 4**. The initial steps involve determining the swarm's parameters, such as its number of populations, positions, velocities, and other relevant features. The objective function that needs to be optimized is also specified. The fitness of each swarm member is then determined by evaluating its present solution based on this objective function. Following evaluation, individuals adjust their positions or solutions, guided by principles inspired by social insects' behavior, such as following pheromone trails or social interactions within the neighborhood. Termination criteria are defined to determine when the algorithm should stop its iteration, such as reaching a maximum number of iterations or achieving a certain fitness level. After the termination requirements are met, the

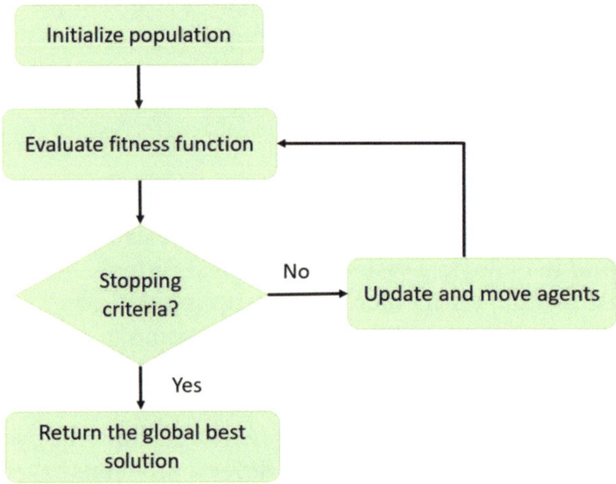

Figure 4.
Steps involve in SI algorithm.

final positions or solutions found by the swarm are analyzed to derive insights from the optimization process.

Swarm-based algorithms balance exploitation (using promising solutions) and exploration (widely searching the solution space). In disaster management, this balance is critical because responders must simultaneously investigate different approaches while exploiting effective solutions promptly.

2.5 Miscellaneous algorithms

Similar algorithms draw inspiration from nature, such as artificial intelligence concepts, game theory, human behavior, mathematics, and political strategies. Teaching-Learning Based Optimization (TLBO) [35], Brain Storm Optimization (BSO) [36], Firefly Algorithm (FA) [37], Football Game Inspired Algorithm (FGIA) [38], Search and Rescue Optimization (SRO) [39] have been utilized in disaster management. Applications such as pattern recognition, network routing, clustering, and decision-making have been tackled by these algorithms. These techniques can be effectively applied to address various optimization challenges and enhance network performance in disaster scenarios.

3. Metaheuristic algorithms for disaster management

Metaheuristic algorithms use in disaster management systems is enumerated in this section. The problems are classified under four major areas as follows.

3.1 Localization

One of the most challenging tasks during or after a disaster scenario is localization of victims. Rescue teams can concentrate their efforts more effectively by reducing response times when the precise location of the victim is known. Effective victim localization minimizes the time and resources required for search and rescue operations, as shown in **Table 1**.

3.2 Routing

In disaster environments, certain characteristics such as signal attenuation and short-term communication links among rescue crews contribute to frequent route disruptions as well as unreliable end-to-end services. Efficient routing in disaster-affected networks maximizes the effective use of scarce resources such as battery power, bandwidth, and processing capacity while guaranteeing reliable data transmission between nodes and central points, such as command center or relief coordination center. **Table 2** shows a summary of papers addressing routing problem.

3.3 Coverage

Coverage extension plays a vital role in disaster management by ensuring that communication networks can reach more affected areas, enhancing the resilience of

Ref.	Year	Types of MA	Description of use
[40]	2019	PSO	Proposed a localization algorithm that combines boundary-based optimization with PSO to locate target UAV nodes efficiently. This methodology achieves lower localization error and higher localization accuracy by analyzing noise parameters that are uniformly distributed with respect to distance and environment.
[41]	2020	PSO	Describes swarm interaction and includes artificial potential functions to support swarm attraction toward victims and collision avoidance, considering the complex and hazardous terrain. The outcomes show how the algorithm allows the rescue team to successfully identify possible victims and avoid collisions.
[42]	2020	GA	A genetic algorithm-based intelligent search system is utilized to identify signal strength variations between distress signals and drones, improving tracking accuracy.
[43]	2021	GWO	Devised a range-free distributed localization algorithm for UAV networks. This method incorporates the bounding cube approach to enhance sampling efficiency, thereby reducing the computational load. This approach reduces distance estimation errors and improves efficiency, facilitating rapid convergence.
[44]	2022	PSO	Proposed a Hierarchical-Reference-PSO to enhance the global search capabilities by evaluating sub-fitness functions of particles. In contrast to traditional PSO, where a combination of local best and global best positions is utilized, HRPSO employs a reference best position to guide particle movements. The suggested approach shows less complexity and localization error, providing a more accurate and efficient solution for UAV localization tasks.
[45]	2023	SSA	The trilateration method is used to estimate the centroid around a specific target node, and Salp swarms are then implemented around the centroid to improve the accuracy of the solution.

Table 1.
Summary of papers for localization problems using MAs.

Ref.	Year	Types of MA	Description of use
[46]	2012	ACO	An ACO optimization is employed to identify the most efficient routes for firefighting and search and rescue missions within buildings. This method and its variations are highly effective at quickly finding the best routes in networks.
[47]	2018	ACO	Introduce a novel ACO-based algorithm that integrates the concept of a virtual central depot to accommodate the multi-depot scenario.
[48]	2019	Squirrel Search Algorithm (SSA)	An enhanced Squirrel Search Algorithm (SSA) for multicast routing is proposed by focusing computation only inside the destination-oriented zone, it effectively minimizes computation time while satisfying constraints such as energy, delay, and packet loss.
[49]	2020	ACO	Devised a hybrid ACO approach where particles serve as operators, expanding the search scope for viable depots and subsequently assigning clients to them. This enables ants to determine the vehicle routes that are most effective and balanced for each selected depot.
[50]	2020	DE	An adaptive selection mutation-constrained DE algorithm is presented to select individuals based on their fitness and constraints and explores the most promising candidates among the selected individuals to guide the mutations. This approach maintains exploration capabilities while improving exploitation.
[51]	2021	PSO	Proposed an enhanced PSO incorporating Variable Neighborhood Search that determines the best next-hop routing depending on factors like distance between nodes and the Base Station (BS) and link reliability, ensuring reliable and timely data transmission to the BS.
[52]	2022	Non-Dominated Sorting Genetic Algorithm-II [NSGA-II], PSO	A mixed integer linear programming (MILP) model is introduced to optimize the routes for emergency vehicles and minimize both the service completion time and the number of patients experiencing worsened conditions due to delayed medical care.
[53]	2023	CSO	Develop an enhanced CSO-based multi-hop routing for choosing the optimal path. This method computes fitness functions to increase network lifetime and energy efficiency using a variety of input variables.

Table 2.
Summary of papers for routing problems using MAs.

communication infrastructure during disasters, and facilitating timely response and recovery actions, as shown in **Table 3**.

3.4 Resource allocation

Disasters often result in widespread damage and disruption, necessitating the rapid mobilization of resources such as SAR operation, equipment, and supplies to affected areas. Effective resource allocation in disaster management is essential for optimizing response efforts, maximizing efficiency, minimizing losses, enhancing coordination and collaboration, and adapting to dynamic conditions as shown in **Table 4**.

Ref.	Year	Types of MA	Description of use
[54]	2012	ACO	Investigates the coverage problem using Ant Colony-Based Scheduling Algorithm. The algorithm uses positional information to assess each sensor's coverage and initializes the pheromone field based on these assessments. Ants prioritize sensors with more coverage, leveraging sensor capabilities to systematically optimize solutions.
[55]	2019	ABC	A UAV-Artificial Bee Colony (U-ABC) algorithm is used for determining the optimal flight positions for each UAV-BS, improves coverage and maximize network throughput within disaster areas.
[56]	2019	PSO	Allows the UAV-BS to find the optimal UAV altitude using PSO by moving straight from its starting position to the ideal position in a single maneuver, improving coverage while consuming less time and energy.
[57]	2020	GOA	Proposed a Grasshopper Optimization Algorithm (GOA)with the aim of maximizing the information derived from assessing areas and roads while simultaneously maximizing total coverage in damaged areas and roads.
[58]	2021	Ant Lion Optimization (ALO)	Presented a coverage enhancement strategy utilizing the Ant Lion Optimization (ALO) algorithm by reallocating antlions and dynamically reducing their numbers and a continual randomly generated boundary shrinkage factor is added.
[59]	2022	PSO	An enhanced PSO model is designed to investigate the most efficient allocation of SAR resources considering various constraints on ship and aircraft base station configurations. The experimental findings indicate that the model effectively optimizes coverage area while allocating maritime rescue resources.
[60]	2023	GWO	Suggested using K-means initialized GWO to deploy 3D ultra-dense small cells (UBS) in order to maximize the number of user equipment served in the disaster area, guaranteeing the fulfillment of data rate requirements for each UE while ensuring that the number of covered UEs does not exceed each UBS's capacity limit.

Table 3.
Summary of papers for coverage problems using MAs.

Ref.	Year	Types of MA	Description of use
[61]	2012	PSO, SA	Introduces an enhanced PSO algorithm for shelter allocation and incorporates SA to improve search efficiency and avoid local optima. In Guangzhou City, China's Zhuguang Block, the suggested algorithm is demonstrated to be effective in allocating earthquake emergency shelters.
[62]	2019	NSGA-II, PSO	NSGA-II and Modified Multiple-Objective PSO, are suggested to minimize the overall cost and the scarcity of relief supplies while optimizing the distribution of facilities. Data from relief bases in the first area of Tehran city are used to validate the proposed model.
[63]	2020	DE, NSGA-II	A Multi-Objective Optimization (MOO) problem is defined to identify optimal SAR strategies and efficiently schedule resource allocation. This methodology enables emergency response quickly and efficiently.
[64]	2020	PSO	Introduces an evolutionary PSO algorithm, incorporating a particle coding scheme and evolutionary operations aligned with the geospatial edge services chain model and presents a fitness evaluation model that considers multiple geospatial edge services characteristics on the edge nodes.

Ref.	Year	Types of MA	Description of use
[65]	2020	GA	Presents a novel approach for post-disaster emergency resource allocation, using GA to address the uncertainties and complexities that occur in rescue operations.
[66]	2021	PSO	PSO and variable neighborhood search are suggested in order to tackle the mixed integer programming model focused on time-cost considerations in the presence of uncertainty.
[67]	2021	PSO	An enhanced PSO algorithm has been introduced, leveraging multi-agent theory and evolutionary population dynamics. This approach optimizes response operations and resource utilization, reducing both time and losses.
[68]	2021	Shuffled Shepherd Optimization (SSO)	Shuffled shepherd optimization (SSO) with a dynamic-window method algorithm is employed to minimize the frequency of recharging or replacement of sensors. This approach considers static and dynamic objects in the environment to prevent collisions and improve energy efficiency during data collection, movements, and transmission.
[69]	2021	DE	While traditional DE relies on random search during the mutation stage, leading to limited exploitation capability, an Exponential Selection-based Differential Evolution (ERDE) introduces exponential selection. This strategy improves the direction of mutation and increases the effectiveness of exploitation by increasing the chance of selecting superior individuals.
[70]	2022	NSGA-II	Multi-objective vibration damping optimization and NSGA-II are used for combined emergency preparedness and response decision-making. The model's objectives are to arrange emergency shelters or depots in the best possible locations and make it easier for relief vehicles to move between disaster sites and these shelters.
[71]	2022	GA	Suggested Elite-preserved Genetic Algorithm (EGA) to optimize the allocation of emergency resources in maritime situations, aiming to enhance efficiency by considering factors such as economic benefits, response time, and safety index.
[72]	2023	JSA	An enhanced Jellyfish Search Algorithm (JSA), which mimics the foraging behavior of jellyfish and is utilized for quick and effective distribution and logistics. During the jellyfish search process, a Cauchy strategy is employed to determine the internal motion's moving distance, enhancing global search capabilities and avoiding local optimal solutions.
[73]	2023	PSO	Proposed a clustering model for optimal target-region allocation using PSO where each affected person is represented as a particle and the fitness function is defined by minimizing the average distance and standard deviation, ensuring that all particles converge to cluster centers by adjusting their positions according to the global best fitness.

Table 4.
Summary of papers for resource allocation problem using MAs.

3.5 Other metaheuristic algorithms

Some new MAs for disaster management are suggested in this section. In recent years, researchers have studied numerous approaches like SRO, TLBO, and Smart Flower Optimization (SFO) and have shown them to be effective and efficient in solving similar kinds of problems, as shown in **Table 5**. These new metaheuristic algorithms still have scope for improvement, making significant enhancements in disaster applications.

Ref.	Year	Types of MA	Description of use
[74]	2017	SRO	Proposed a novel Swarm Robotics SRO using robots with environmental awareness for search and rescue missions in disaster sites. The locations of the survivors act as the global best solutions, with one robot distinguished as the master and others as slaves. The outcomes show improved convergence speed and quality of the solutions.
[75]	2021	SRO	Introduced a novel Chaotic SRO-based multi-hop data transmission by incorporating chaotic principles into the conventional SRO algorithm, designed to enhance energy efficiency and prolong the network lifetime.
[76]	2022	PSO	An innovative approach called Reward-based Reinforcement Learning integrating Particle Swarm Optimization (R-RLPSO) is proposed for real-time rescue operations in 3-D underwater environments and provides cost-effectiveness and quick response.
[77]	2022	SFO	Presented a new MA for efficient UAV path planning in a shorter period of time. Two growth strategies are used in this approach to control the movement of immature flowers, and particle positions are updated through mathematical modeling.
[78]	2023	TLBO	A discrete-TLBO is employed to minimize the average completion time for all SAR teams. The algorithm creates a two-phase local search method to enhance the solution after producing an initial task scheduling sequence for every rescue team.
[79]	2023	SRO	Designed a heuristic crossover approach with the fundamental SRO technique to enhance convergence speed while preserving population diversity during the optimization procedure, ensuring high convergence speed and enhancing path planning effectiveness.
[80]	2023	ACO	Introduces an improved ACO algorithm to enhance the discovery of a mobile search object exhibiting Markovian motion, optimizing search success within a short time and limited resources. This improved algorithm achieves search trajectories with higher success rates in shorter times.

Table 5.
Summary of papers using new MAs for disaster management.

4. Hybrid metaheuristic algorithms

A hybrid optimization algorithm is a technique that combines multiple optimization techniques to solve a given problem. By leveraging the features of different algorithms, hybrid optimization aims to improve solution quality, convergence speed, and robustness, particularly for complex and challenging optimization problems. **Table 6** provides a summary of papers using hybrid MAs.

5. Limitations and discussion

According to the results of the literature review, MAs are still in the early stages of development in tackling post-disaster problems like localization of victims and coverage extension problems. There has been limited research explored to overcome these challenges. MAs often require careful parameter tuning to achieve optimal performance, a process that can be both time-consuming and problem-specific process. Thus, future research should prioritize long-term disaster recovery, leveraging hybridization and machine learning algorithms.

Ref.	Year	Types of MA	Description of use
[81]	2014	Biogeography-Based Optimization (BBO), DE	A biogeography-based optimization (BBO) algorithm is integrated with the DE mutation operator to improve exploration capabilities and adopt a local ring topology within the population to prevent premature convergence. This approach reduces the weighted time required to deliver all necessary supplies to the target area.
[82]	2016	DE, PSO	A multi-objective hybrid DE and PSO algorithm is proposed to generate a set of Pareto solutions aimed at minimizing the use of rescue vehicles as well as the extinguishing time in combating forest fires, considering limited resources.
[83]	2018	AIS, ACO	A hybrid algorithm is designed to address the routing problem of emergency vehicle scheduling for grain transportation by assessing crowd density and performs non-dominated sorting of the population in the ACO phase, minimizing the distribution time of vehicle scheduling.
[84]	2018	NSGA-II, ACO	Employs a hybrid algorithm to investigate the emergency rescue vehicle dispatch problem in order to generate more effective initial solutions that minimize rescue times, delay costs, and maximize lifesaving utility.
[85]	2019	PSO, HSA	Devised an enhanced hybrid method for selecting cluster heads and relay nodes, incorporating a novel fitness function that considers energy efficiency and communication link quality. This approach extends network coverage and network lifetime by optimizing multi-hop routing.
[86]	2022	BSO, SA	Addresses the routing problem by integrating the Brain Storm Optimization (BSO) algorithm with Large Neighborhood Search (LNS) and simulated SA techniques. BSO generates an initial solution through K-means clustering, where a cluster center is chosen randomly and updated. These solutions are further optimized using LNS, which involves splitting and regrouping individuals using "destroy" and "repair" operators to improve their quality.
[87]	2023	GWO, DE	Designed a hybrid algorithm to maximize GWO's exploitation potential where the alpha wolf explores its surroundings, with the beta and delta wolves closely following the alpha's movements. Moreover, DE gains from a strategically integrated rank-based mutation technique that increases its potential for exploitation.

Table 6.
Summary of papers using hybrid MAs for disaster management.

The common MAs, such as Genetic Algorithm (GA), Simulated Annealing (SA), Particle Swarm Optimization (PSO), and Ant Colony Optimization (ACO), employ strategies that allow for comprehensive and more efficient exploration of solution spaces. These algorithms use mechanisms like randomization and probabilistic decision-making to escape local optima and cover more solution space quickly. GA and PSO can optimize the logistics of resource distribution by finding the best routes and schedules, considering constraints like road conditions and supply availability, and thereby minimizing cost and time. MAs, including ACO and SA, are effective in solving network routing problems and evacuation route planning considering dynamic factors. Some research has developed hybrid approaches by combining various metaheuristic approaches in order to provide better solutions.

6. Benefits and challenges

Metaheuristic algorithms offer several benefits when applied to disaster scenarios, leading to more effectiveness and efficiency in disaster response and recovery efforts. MAs algorithms can efficiently explore large solution spaces and find near-optimal solutions within reasonable timeframes. This efficiency is crucial in time-sensitive disaster situations where quick decision-making is necessary to save lives and minimize damage. Moreover, MAs can handle noisy, incomplete, or uncertain parameters in disaster scenarios, ensuring reliable decision-making even in challenging conditions. MAs can handle MOO problems by generating a set of Pareto-optimal solutions that represent trade-offs between competing objectives, enabling decision-makers to explore and select the most suitable solutions based on their preferences and priorities.

While metaheuristic algorithms offer promising solutions for optimization problems in disaster management, it also faces several challenges. MAs generally have several parameters that need to be tuned to achieve optimal performance. Finding the right parameter settings for a specific disaster management problem can be challenging and may involve extensive experimentation. Many disaster management problems involve multiple conflicting objectives, such as minimizing cost, maximizing resource utilization, reducing energy, and maximizing coverage. MAs need to balance these conflicting objectives and find a set of solutions that represent trade-offs between them, which can be challenging.

MAs often rely on data to inform decision-making processes. During disaster scenarios, data may be limited, incomplete, or of poor quality due to infrastructure damage, communication breakdowns, or other challenges. As a result, the effectiveness of MAs can be compromised, leading to suboptimal solutions. Moreover, most computational experiments are usually performed on test problems that are generated at random; very few studies use real-world data for testing.

Quick decisions must be made during emergency situations. In order to reduce the effects of disasters and save lives, decision-makers frequently need to act quickly and find solutions in real time. However, many MAs require considerable computational time to converge to a solution, making them less suitable for time-critical decision-making. Different metaheuristic algorithms have varying levels of computational complexity. Some metaheuristic algorithms have complex operations and require a significant number of iterations to converge to adequate solutions. This complexity contributes to higher computational costs, especially when dealing with high-dimensional optimization problems or problems with complex constraints.

7. Conclusion and future perspectives

7.1 Summary of key findings

Natural disasters pose significant causes of human lives and damage to infrastructure and property. The application of metaheuristic algorithms has increasingly addressed the complexity associated with managing these disasters. This study discusses an overview of recent research developments in metaheuristic optimization for disaster management. We have discussed various emergency and disaster relief problems using MAs and hybrid optimization techniques. Recent research on the utilization of MAs for disaster management offers faster and more effective solutions.

Compared to conventional methods, MAs are more effective at quickly exploring vast solution spaces, have inherent adaptability to uncertainty and randomness, are robust against noise, and have scalability to high dimensions. Therefore, they are highly suitable for addressing emergency and disaster relief operations. This study shows the key findings and benefits of MAs in enhancing network coverage, optimizing resource utilization, and minimizing response time and cost in emergency and disaster relief operations.

7.2 Future research directions

Considering the challenges, the following aspects highlight the future development of MAs in disaster management:

- *Hybrid and adaptive algorithms:* Researchers can investigate the development of hybrid optimization tools that combine metaheuristic algorithms with artificial intelligence, aiming to improve solutions and reduce computation time by effectively balancing exploration and exploitation. Hybrid algorithms have proven effective in solving a wide range of planning and scheduling problems, but their potential for use in emergency optimization issues remains to be explored.

- *Real-world applications and case studies*: There are currently very few publicly available benchmark problems in this field. More standard test suites and advanced simulation tools for evaluating and analyzing different MAs must be developed. In order to validate the effectiveness of these algorithms, more case studies and real-world applications should be conducted and using feedback loop mechanism from real-world implementations to continuously improve and adapt the algorithms.

- *Multi-objective optimization and decision support:* The utilization of multi-objective optimization techniques is expected to provide significant benefits for studies focused on optimizing disaster relief operations. Utilizing Pareto fronts to illustrate trade-offs between different objectives so that decision-makers can choose the best-balanced solutions.

The study has identified research gaps and suggests potential future research directions. Additionally, it highlights the necessity for further real-world applications to validate the applicability of these proposed approaches in uncertain emergency scenarios.

Conflict of interest

The authors state that they have no known competing financial interests or personal relationships that could have appeared to influence the work reported in this paper.

DOI: http://dx.doi.org/10.5772/intechopen.1006175

Author details

Bidyarani Langpoklakpam and Lithungo K. Murry*
Department of Computer Science and Engineering, National Institute of Technology, Dimapur, Nagaland, India

*Address all correspondence to: lithungo@nitnagaland.ac.in

References

[1] Debnath S, Arif W, Roy S, Baishya S, Sen D. A comprehensive survey of emergency communication network and management. Wireless Personal Communications. 2022;**124**(2):1375-1421

[2] Wang Q, Li W, Yu Z, Abbasi Q, Imran M, Ansari S, et al. An overview of emergency communication networks. Remote Sensing. 2023;**15**(6):1595

[3] Damaševičius R. From sensors to safety: Internet of emergency services (IoES) for emergency response and disaster management. Journal of Sensor and Actuator Networks. 2023;**12**(3):41

[4] Akwafuo SE, Mikler AR, Irany FA. Optimization models for emergency response and post-disaster delivery logistics: A review of current approaches. International Journal of Engineering Technologies and Management Research. 2020;**7**(8):35-49

[5] Bayram V. Optimization models for large scale network evacuation planning and management: A literature review. Surveys in Operations Research and Management Science. 2016;**21**(2):63-84

[6] Abualigah L, Elaziz MA, Khasawneh AM, Alshinwan M, Ibrahim RA, Al-Qaness MA, et al. Meta-heuristic optimization algorithms for solving real-world mechanical engineering design problems: A comprehensive survey, applications, comparative analysis, and results. Neural Computing and Applications. 2022;**30**:1-30

[7] Rajwar K, Deep K, Das S. An exhaustive review of the metaheuristic algorithms for search and optimization: Taxonomy, applications, and open challenges. Artificial Intelligence Review. 2023;**56**(11):13187-13257

[8] Slowik A, Kwasnicka H. Evolutionary algorithms and their applications to engineering problems. Neural Computing and Applications. 2020;**32**:12363-12379

[9] Srinivas M, Patnaik LM. Genetic algorithms: A survey. Computer. 1994;**27**(6):17-26

[10] Das S, Suganthan PN. Differential evolution: A survey of the state-of-the-art. IEEE Transactions on Evolutionary Computation. 2010;**15**(1):4-31

[11] Li Z, Lin X, Zhang Q, Liu H. Evolution strategies for continuous optimization: A survey of the state-of-the-art. Swarm and Evolutionary Computation. 2020;**56**:100694

[12] Fogel DB. Evolutionary programming: An introduction and some current directions. Statistics and Computing. 1994;**4**:113-129

[13] Espejo PG, Ventura S, Herrera F. A survey on the application of genetic programming to classification. IEEE Transactions on Systems, Man, and Cybernetics. 2009;**40**(2):121-144

[14] Su H, Zhao D, Heidari AA, Liu L, Zhang X, Mafarja M, et al. RIME: A physics-based optimization. Neurocomputing. 2023;**532**:183-214

[15] Alia OMD, Mandava R. The variants of the harmony search algorithm: An overview. Artificial Intelligence Review. 2011;**36**:49-68

[16] Bertsimas D, Tsitsiklis J. Simulated annealing. Statistical Science. 1993;**8**(1):10-15

[17] Rashedi E, Rashedi E, Nezamabadi-Pour H. A comprehensive survey on

gravitational search algorithm. Swarm and Evolutionary Computation. 2018;**41**:141-158

[18] Abualigah L, Elaziz MA, Sumari P, Khasawneh AM, Alshinwan M, Mirjalili S, et al. Black hole algorithm: A comprehensive survey. Applied Intelligence. 2022;**52**(10):11892-11915

[19] Neri F, Cotta C. Memetic algorithms and memetic computing optimization: A literature review. Swarm and Evolutionary Computation. 2012;**2**:1-14

[20] Maaroof BB, Rashid TA, Abdulla JM, Hassan BA, Alsadoon A, Mohammadi M, et al. Current studies and applications of shuffled frog leaping algorithm: A review. Archives of Computational Methods in Engineering. 2022;**29**(5):3459-3474

[21] Sánchez-García J, García-Campos JM, Reina DG, Toral SL, Barrero F. Application of nature inspired algorithms for wireless multi-hop ad hoc network optimization problems in disaster response scenarios. In: Nature-Inspired Networking. 1st ed. Florida, Boca Raton; 2018. pp. 49-88

[22] Shehab M, Abu-Hashem MA, Shambour MKY, Alsalibi AI, Alomari OA, Gupta JN, et al. A comprehensive review of bat inspired algorithm: Variants, applications, and hybridization. Archives of Computational Methods in Engineering. 2023;**30**(2):765-797

[23] Shehab M, Khader AT, Al-Betar MA. A survey on applications and variants of the cuckoo search algorithm. Applied Soft Computing. 2017;**61**:1041-1059

[24] Panda M, Das B. Grey wolf optimizer and its applications: A survey. In: Proceedings of the Third International Conference on Microelectronics, Computing and Communication Systems. Lecture Notes in Electrical Engineering. Vol. 556. Springer Singapore; 2018. pp. 179-194

[25] Abdel-Basset M, Shawky LA. Flower pollination algorithm: A comprehensive review. Artificial Intelligence Review. 2019;**52**:2533-2557

[26] Mirjalili S, Lewis A. The whale optimization algorithm. Advances in Engineering Software. 2016;**95**:51-67

[27] Shopon M, Adnan MA, Mridha MF. Krill herd based clustering algorithm for wireless sensor networks. In: 2016 International Workshop on Computational Intelligence (IWCI); Dhaka, Bangladesh: IEEE; 2016. pp. 96-100. DOI: 10.1109/IWCI.2016.7860346

[28] Timmis J, Knight T, de Castro LN, Hart E. An overview of artificial immune systems. In: Computation in Cells and Tissues. Natural Computing Series. Berlin, Heidelberg: Springer; 2004. pp. 51-91. DOI: 10.1007/978-3-662-06369-9_4

[29] Kameyama K. Particle swarm optimization-a survey. IEICE Transactions on Information and Systems. 2009;**92**(7):1354-1361

[30] Dorigo M, Blum C. Ant colony optimization theory: A survey. Theoretical Computer Science. 2005;**344**(2-3):243-278

[31] Kaipa KN, Ghose D. Glowworm swarm optimization: Algorithm development. In: Glowworm Swarm Optimization. Studies in Computational Intelligence. Vol. 698. Cham: Springer; 2017. pp. 21-56. DOI: 10.1007/978-3-319-51595-3_2

[32] Bansal JC, Sharma H, Jadon SS. Artificial bee colony algorithm: A

survey. International Journal of Advanced Intelligence Paradigms. 2013;**5**(1-2):123-159

[33] Neshat M, Sepidnam G, Sargolzaei M, Toosi AN. Artificial fish swarm algorithm: A survey of the state-of-the-art, hybridization, combinatorial and indicative applications. Artificial Intelligence Review. 2014;**42**(4):965-997

[34] Hussien AG, Amin M, Wang M, Liang G, Alsanad A, Gumaei A, et al. Crow search algorithm: Theory, recent advances, and applications. IEEE Access. 2020;**8**:173548-173565

[35] Rao RV, Savsani VJ, Vakharia DP. Teaching–learning-based optimization: A novel method for constrained mechanical design optimization problems. Computer-aided Design. 2011;**43**(3):303-315

[36] Cheng S, Qin Q, Chen J, Shi Y. Brain storm optimization algorithm: A review. Artificial Intelligence Review. 2016;**46**:445-458

[37] Johari NF, Zain AM, Noorfa MH, Udin A. Firefly algorithm for optimization problem. Applied Mechanics and Materials. 2013;**421**:512-517

[38] Fadakar E, Ebrahimi M. A new metaheuristic football game inspired algorithm. In: 2016 1st Conference on Swarm Intelligence and Evolutionary Computation (CSIEC). Bam, Iran: IEEE; 2016. pp. 6-11. DOI: 10.1109/CSIEC.2016.7482120

[39] Shabani A, Asgarian B, Salido M, Gharebaghi SA. Search and rescue optimization algorithm: A new optimization method for solving constrained engineering optimization problems. Expert Systems with Applications. 2020;**161**:113698

[40] Arafat MY, Moh S. Localization and clustering based on swarm intelligence in UAV networks for emergency communications. IEEE Internet of Things Journal. 2019;**6**(5):8958-8976

[41] Paez D. Distributed particle swarm optimization for multi-robot system in search and rescue operations. IFAC-PapersOnLine. 2020;**54**(4):1-6

[42] Oh D, Han J. Smart search system of autonomous flight UAVs for disaster rescue. Sensor. 2020;**21**(20):6810

[43] Arafat MY, Moh S. Bio-inspired approaches for energy-efficient localization and clustering in UAV networks for monitoring wildfires in remote areas. IEEE. Access. 2021;**6**:18649-18669

[44] Zhang W, Zhang W. An efficient UAV localization technique based on particle swarm optimization. IEEE Transactions on Vehicular Technology. 2022;**71**(9):9544-9557

[45] Rani S, Babbar H, Kaur P, Alshehri MD, Shah SH. An optimized approach of dynamic target nodes in wireless sensor network using bio inspired algorithms for maritime rescue. IEEE Transactions on Intelligent Transportation Systems. 2023;**24**(2):2548-2555

[46] Wu C, Chen L. 3D spatial information for fire-fighting search and rescue route analysis within buildings. Fire Safety Journal. 2012;**48**:21-29

[47] Wang X, Choi T-M, Liu H, Yue X. A novel hybrid ant colony optimization algorithm for emergency transportation problems during post-disaster scenarios. IEEE Transactions on Systems, Man, and Cybernetics: Systems. 2018;**48**(4):545-556

[48] Chaudhry R, Tapaswi S, Kumar N. A green multicast routing algorithm for smart sensor networks in disaster management. IEEE Transactions on Green Communications and Networking. 2019;**3**(1):215-226

[49] Wei X. An integrated location-routing problem with post-disaster relief distribution. Computers & Industrial Engineering. 2020;**147**:106632

[50] Yu X. A constrained differential evolution algorithm to solve UAV path planning in disaster scenarios. Knowledge-Based Systems. 2020;**204**:106209

[51] Sun G, Qin D, Lan T, Ma L. Research on clustering routing protocol based on improved PSO in FANET. IEEE Sensors Journal. 2021;**21**(23):27168-27185

[52] Rabbani M, Oladzad-Abbasabady N, Akbarian-Saravi N. Ambulance routing in disaster response considering variable patient condition: NSGA-II and MOPSO algorithms. Journal of Industrial & Management Optimization. 2022;**18**(2):1035

[53] Santhanaraj RK, Rajendran S, Romero CAT, Murugaraj SS. Internet of things enabled energy aware metaheuristic clustering for real time disaster management. Computer Systems Science & Engineering. 2023;**45**(2):1561-1576

[54] Lee J-W, Lee J-J. Ant-colony-based scheduling algorithm for energy-efficient coverage of WSN. IEEE Sensors Journal. 2012;**12**(10):3036-3046

[55] Li J, Lu D, Zhang G, Tian J, Pang Y. Post-disaster unmanned aerial vehicle base station deployment method based on artificial bee colony algorithm. IEEE Access. 2019;**7**:168327-168336

[56] Hydher H. Intelligent UAV deployment for a disaster-resilient wireless network. Sensors. 2019;**20**(21):6140

[57] Danesh Alagheh Band TS, Aghsami A, Rabbani M. A post-disaster assessment routing multi-objective problem under uncertain parameters. International Journal of Engineering. 2020;**33**(12):2503-2508

[58] Yao Y, Li Y, Xie D, Hu S, Wang C, Li Y. Coverage enhancement strategy for WSNs based on virtual force-directed ant lion optimization algorithm. IEEE Sensors Journal. 2021;**21**(17):19611-19622

[59] Sun Y. Exploring maritime search and rescue resource allocation via an enhanced particle swarm optimization method. Journal of Marine Science and Engineering. 2022;**10**(7):906

[60] Mandloi D, Arya R. QoS-aware UAV mounted base station deployment in a disaster scenario with user position estimation. E-Prime - Advances in Electrical Engineering, Electronics and Energy. 2023;**6**:100281

[61] Hu F, Wei X, Li X. A modified particle swarm optimization algorithm for optimal allocation of earthquake emergency shelters. International Journal of Geographical Information Science. 2012;**26**(9):1643-1666

[62] Ghasemi P. Uncertain multi-objective multi-commodity multi-period multi-vehicle location-allocation model for earthquake evacuation planning. Applied Mathematics and Computation. 2019;**350**:105-132

[63] Xiong W. A decision support method for design and operationalization of search and rescue in maritime emergency. Ocean Engineering. 2020;**207**:107399

[64] Liao JX, Wu XW. Resource allocation and task scheduling scheme

in priority-based hierarchical edge computing system. In: 2020 19th International Symposium on Distributed Computing and Applications for Business Engineering and Science (DCABES). Xuzhou, China: IEEE; 2020. pp. 46-49. DOI: 10.1109/DCABES50732.2020.00021

[65] Wang F, Pei Z, Dong L, Ma. Emergency resource allocation for multi-period post-disaster using multi-objective cellular genetic algorithm. IEEE Access. 2020;**8**:82255-82265

[66] Wang BC, Qian QY, Gao JJ, Tan ZY, Zhou Y. The optimization of warehouse location and resources distribution for emergency rescue under uncertainty. Advanced Engineering Informatics. 2021;**48**:101278

[67] Ye X, Chen B, Lee K. An emergency response system by dynamic simulation and enhanced particle swarm optimization and application for a marine oil spill accident. Journal of Cleaner Production. 2021;**297**:126591

[68] Khan A, Zhang J, Ahmad S, Memon S, Qureshi HA, Ishfaq M. Dynamic positioning and energy-efficient path planning for disaster scenarios in 5G-assisted multi-UAV environments. Electronics. 2021;**11**(14):2197

[69] Zhang X, Yu X, Wu X. Exponential rank differential evolution algorithm for disaster emergency vehicle path planning. IEEE Access. 2021;**9**:10880-10892

[70] Seraji H, Tavakkoli-Moghaddam R. Asian: An integrative location-allocation model for humanitarian logistics with distributive injustice and dissatisfaction under uncertainty. Annals of Operations Research. 2022;**319**:211-257

[71] Ma Q, Zhang D, Wan C, Zhang J, Lyu N. Multi-objective emergency resources allocation optimization for maritime search and rescue considering accident black-spots. Ocean Engineering. 2022;**261**:112178

[72] Li P, Fan X. The application of the improved jellyfish search algorithm in a site selection model of an emergency logistics distribution center considering time satisfaction. Biomimetics. 2023;**8**(4):349

[73] Li J, Cao S, Liu X, Ruiyun Y, Wang X. PSO and MADDPG based multi-UAVs trajectory planning algorithm for emergency communication. Frontiers in Neurorobotics. 2023;**16**:1076338

[74] Bakhshipour M, Jabbari Ghadi M, Namdari F. Swarm robotics search & rescue: A novel artificial intelligence-inspired optimization approach. Applied Soft Computing. 2017;**57**:708-726

[75] Anuradha D, Subramani N, Khalaf OI, Alotaibi Y, Alghamdi S, Rajagopal M. Chaotic search-and-rescue-optimization-based multi-hop data transmission protocol for underwater wireless sensor networks. Sensors. 2021;**22**(8):2867

[76] Wu J, Song C, Ma J, Wu J, Han G. Reinforcement learning and particle swarm optimization supporting real-time rescue assignments for multiple autonomous underwater vehicles. IEEE Transactions on Intelligent Transportation Systems. 2022;**23**(7):6807-6820

[77] Qadir Z, Zafar MH, Moosavi SKR, Le KN, Tam VW. Optimizing UAV path for disaster management in smart cities using metaheuristic algorithms. In: Computational Intelligence for Unmanned Aerial Vehicles Communication Networks. Vol. 1033. Cham: Springer International

Publishing; 2022. pp. 225-244. DOI: 10.1007/978-3-030-97113-7_13

[78] Xu Y, Li X, Li Q. A discrete teaching–learning based optimization algorithm with local search for rescue task allocation and scheduling. Applied Soft Computing. 2023;**134**:109980

[79] Zhang C. A novel UAV path planning approach: Heuristic crossing search and rescue optimization algorithm. Expert Systems with Applications. 2023;**215**:119243

[80] Morin M, Abi-Zeid I, Quimper C-G. Ant colony optimization for path planning in search and rescue operations. European Journal of Operational Research. 2023;**305**(1):53-63

[81] Zheng Y-J, Ling H-F, Shi H-H, Chen H-S, Chen S-Y. Emergency railway wagon scheduling by hybrid biogeography-based optimization. Computers & Operations Research. 2014;**43**:1-8

[82] Tian G, Ren Y, Zhou M. Dual-objective scheduling of rescue vehicles to distinguish forest fires via differential evolution and particle swarm optimization combined algorithm. IEEE Transactions on Intelligent Transportation Systems. 2016;**17**(11):3009-3021

[83] Zhang Q, Xiong S. Routing optimization of emergency grain distribution vehicles using the immune ant colony optimization algorithm. Applied Soft Computing. 2018;**71**:917-925

[84] Cao J, Han H, Jiang Y-P, Wang Y-J. Emergency rescue vehicle dispatch planning using a hybrid algorithm. International Journal of Information Technology & Decision Making. 2018;**17**(06):1865-1890

[85] Biabani M, Fotouhi H, Yazdani N. An energy-efficient evolutionary clustering technique for disaster management in IoT networks. Sensors. 2019;**20**(9):2647

[86] Wang X. A hybrid brain storm optimization algorithm to solve the emergency relief routing model. Sustainability. 2022;**15**(10):8187

[87] Yu X. A hybrid algorithm based on grey wolf optimizer and differential evolution for UAV path planning. Expert Systems with Applications. 2023;**215**:119327

Chapter 6

MDWT: A Modified Discrete Wavelet Transformation-Based Algorithm for Image Fusion

Gargi J. Trivedi

Abstract

In this paper, we propose a novel method for image fusion that utilizes a modified version of the discrete wavelet transform (DWT). The proposed algorithm enhances the standard DWT-based techniques by introducing several modifications to the traditional DWT algorithm. We use the modified DWT algorithm to decompose the source images into different sub-bands, which are then combined using a fusion strategy to produce a high-quality fused image. Experimental results demonstrate that our proposed algorithm significantly outperforms traditional DWT-based and PCA-based methods, achieving higher average values in key image quality metrics such as peak signal-to-noise ratio (PSNR), structural similarity index (SSIM), and feature similarity index (FSIM). Specifically, the MDWT method achieved an average PSNR of 30.8 dB, SSIM of 0.880, and FSIM of 0.860, indicating superior image quality. Our study highlights the potential of the modified DWT algorithm to enhance the performance of image fusion techniques, making it an attractive option for researchers and practitioners in the field.

Keywords: image fusion, discrete wavelet transform (DWT), modified discrete wavelet transform (MDWT), fusion strategy, transform domain fusion, multimodal Image processing

1. Introduction

Image fusion is the process of combining two or more images of the same scene, acquired from different sensors or modalities, into a single composite image that provides more information than the individual images alone [1–3]. It has a wide range of applications in various fields, such as remote sensing, medical imaging, surveillance, and robotics. One of the most popular techniques for image fusion is based on the discrete wavelet transform (DWT) [4].

In recent years, significant research has focused on DWT-based image fusion techniques due to their effectiveness. However, traditional DWT-based methods suffer from limitations such as shift variance and poor directional selectivity. To address these issues, researchers have proposed various modifications to the DWT algorithm. For example, the redundant DWT (RDWT) reduces shift variance by adding redundant wavelet coefficients [5], and the dual-tree complex wavelet transform (DT-CWT) improves directional selectivity by using two DWTs with different orientations [6–8].

Additionally, different fusion strategies have been explored to enhance DWT-based image fusion. The pulse-coupled neural network (PCNN) is one such strategy that enhances the saliency of the fused image by simulating neuronal behavior in the visual cortex [9]. Despite these advances, traditional DWT-based methods remain widely used, indicating a need for further improvements in accuracy and efficiency.

Recent advancements in biomedical image fusion have significantly enhanced the accuracy and effectiveness of diagnostic and therapeutic processes [10–13]. One notable State of the Art (SOTA) approach is the use of explainable deep learning for diagnosing Alzheimer's disease [14, 15]. In this method, researchers utilize positron emission tomography (PET) and magnetic resonance imaging (MRI) to generate fused images that provide a comprehensive view of the brain's structure and function. The integration of PET and MRI images allows for a more accurate detection of Alzheimer's disease by highlighting both the metabolic changes and anatomical features of the brain. This approach not only improves diagnostic accuracy but also provides interpretability, which is crucial for clinical decision-making [16].

Another innovative approach in biomedical image fusion involves the use of pixel-level fusion with vision transformers for early detection of Alzheimer's disease [17, 18]. Vision transformers, which have shown remarkable performance in various image processing tasks, are employed to fuse images at the pixel level, capturing intricate details and correlations between different modalities [19]. This method enhances the precision of the fused images, leading to improved early detection of Alzheimer's disease. The ability to detect subtle changes in the brain at an early stage is critical for timely intervention and treatment, making this approach highly valuable in clinical settings [16, 20].

Additionally, Pareto optimized adaptive learning with transposed convolution has been explored for the classification of Alzheimer's disease. This method optimizes the fusion process by balancing multiple objectives, such as accuracy and computational efficiency, through adaptive learning techniques. The use of transposed convolution allows for the generation of high-resolution fused images, which are essential for detailed analysis and accurate classification. This approach demonstrates the potential of advanced machine learning techniques in enhancing biomedical image fusion, providing robust tools for the diagnosis and monitoring of Alzheimer's disease and other neurological conditions [10, 21].

These SOTA techniques highlight the ongoing advancements in biomedical image fusion, emphasizing the importance of integrating multiple imaging modalities and leveraging advanced machine learning algorithms to improve diagnostic accuracy and clinical outcomes [22, 23]. The continued development of these innovative methods is expected to play a crucial role in advancing the field of medical imaging and healthcare [24, 25].

In this paper, we propose a novel algorithm for image fusion using a modified version of the discrete wavelet transform (DWT) that addresses the limitations of traditional DWT-based methods. Our algorithm improves directional selectivity and reduces shift variance, providing a more accurate representation of the source images. We employ a fusion strategy based on the weighted average concept, which enhances the performance of image fusion algorithms. Experimental results demonstrate that our proposed algorithm achieves superior visual quality and objective evaluation metrics compared to other state-of-the-art image fusion techniques.

The paper is organized as follows: the Introduction section provides an overview of the proposed method for image fusion and its potential applications. The Materials and Methods section describes the modified DWT algorithm, including enhancements for improved directional selectivity and reduced shift variance, and explains the fusion strategy used to combine the sub-bands. The Experimental Analysis section covers key parameters, selection criteria, the influence of parameters, and system information, followed by a quantitative analysis and statistical analysis of the experimental results. The Results and Discussion section presents the experimental results and compares the proposed method's performance with other state-of-the-art techniques using objective evaluation metrics. Finally, the Conclusions section summarizes the key findings, highlights the potential of the modified DWT algorithm for advancing image fusion techniques, and discusses future research directions.

2. Methods

To achieve high-quality image fusion, various mathematical and computational techniques have been employed. Principal component analysis (PCA) and the discrete wavelet transform (DWT) are two widely used approaches due to their capability to preserve significant features of the source images. However, these methods have limitations in handling complex structures or achieving optimal performance metrics. In this study, we introduce a modified discrete wavelet transform (MDWT), an enhancement to the traditional DWT, to overcome these challenges and improve fusion quality.

2.1 Principal component analysis (PCA) for image fusion

Principal component analysis (PCA) is a statistical method used to transform a set of correlated variables into a set of uncorrelated variables called principal components [26]. For image fusion, PCA is utilized to combine the information from multiple images into a single image that retains the essential features of the input images. Initially, the input images are converted into column vectors and concatenated to form a data matrix. The covariance matrix of this data matrix is then computed to understand the variance and correlation between different images. The eigenvalues and eigenvectors of the covariance matrix are calculated, where the eigenvectors represent the principal components, and the eigenvalues indicate the amount of variance captured by each principal component. The principal components are sorted based on their eigenvalues in descending order, and the top components that capture the most variance are selected. Finally, these selected principal components are combined to form the fused image, ensuring that it contains significant information from the input images. This process effectively reduces the dimensionality of the data while preserving the most significant features, making PCA a powerful tool for image fusion applications where enhanced image quality and information content are critical [2, 27].

2.2 Discrete wavelet transform

The discrete wavelet transform (DWT) is a widely used technique for signal and image processing [2, 4]. It decomposes a signal or image into different frequency bands, allowing for efficient compression, noise reduction, and feature extraction. The DWT operates by recursively dividing the signal into two halves and then applying wavelet filters to each half, resulting in two sets of coefficients: approximation coefficients (low-frequency components) and detail coefficients (high-frequency components) [25].

$$\begin{cases} c_{j+1,k} = \sum_{n=0}^{N-1} h_n \cdot c_{(j,k\times 2+n)} \\ d_{j+1,k} = \sum_{n=0}^{N-1} g_n \cdot C_{j,k\times 2+n} \end{cases} \tag{1}$$

where $c_{(j,k)}$ represents the approximation coefficients at level j and position k, $d_{(j,k)}$ represents the detail coefficients at level j and position k, h_n and g_n represent the scaling and wavelet coefficients, respectively, and N is the length of the filters.

The inverse DWT can be used to reconstruct the original signal from the approximation and detail coefficients by convolving the coefficients with the inverse wavelet filters.

2.3 Modified discrete wavelet transformation

The proposed modified discrete wavelet transformation (MDWT) algorithm introduces modifications to the traditional DWT to enhance directional selectivity and reduce shift variance. This involves using a different wavelet function and incorporating a novel fusion strategy.

Here is a step-by-step algorithm for the proposed method of image fusion using modified DWT (**Figure 1**).

2.3.1 Input

Two input images, I_1 and I_2 which are to be fused.

2.3.1.1 Preprocessing

Before MDWT decomposition, the input images are preprocessed to enhance their quality. This includes image resizing and histogram equalization.

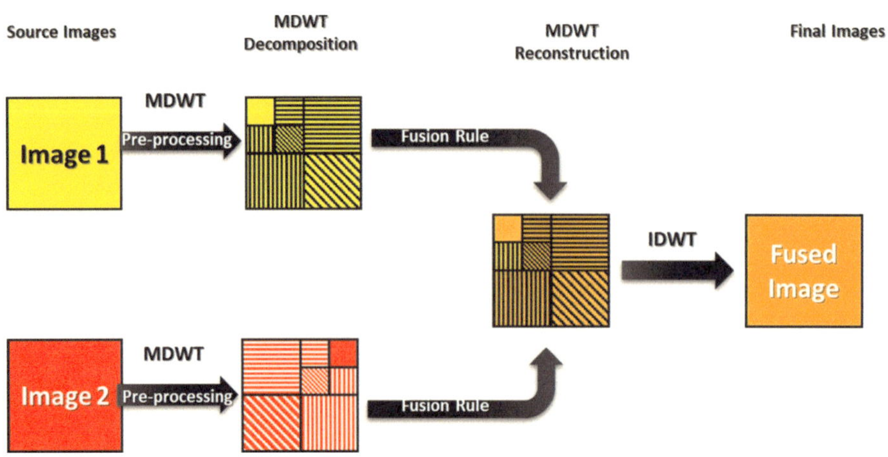

Figure 1.
Block diagram of modified discrete wavelet transformation (MDWT).

2.3.1.2 Image resizing

Ensures the two input images are of the same size and resolution.

$I_{out} = \text{imresize}(I_{in}, [M, N])$ where I I_{in} is the input image, I_{out} is the output image after resizing, and I_{out} is the desired size of the output image.

2.3.1.3 Histogram equalization

It is performed to enhance the contrast of the input images. This is done by stretching the intensity values of the input images to cover the full dynamic range of the display device. The stretching function is applied to the pixel values of the input images using the following mathematical equation:

$$I_{out}(i, j) = \frac{I_{in}(i, j) - \min(I_{in})}{\max(I_{in}) - \min I_{in}} \times 255 \tag{2}$$

where $I_{in}(i, j)$ is the pixel value of the input image at position (i, j) before histogram equalization, $I_{out}(i, j)$ is the pixel value of the input image at position (i, j) after histogram equalization, and $\min(I_{in})$ and $\max(I_{in})$ are the minimum and maximum pixel values of the input image, respectively. The output of the preprocessing step is two preprocessed input images that are of the same size and resolution and have enhanced contrast. These preprocessed input images are then used as inputs to the modified DWT algorithm for image fusion.

2.3.2 MDWT decomposition

Preprocessed input images are decomposed using a modified DWT algorithm that utilizes a specific wavelet function. Using following equation.

$$\begin{cases} W_{j,k} = \frac{1}{\sqrt{2}} \left[H_k^j \cdot I_j^L + G_k^j \cdot I_j^H \right] \\ V_{j+1,k} = \frac{1}{\sqrt{2}} \left[H_k^j \cdot I_j^L - G_k^j \cdot I_j^H \right] \end{cases} \tag{3}$$

where I_j^L and I_j^H are the low-pass and high-pass components of the input signal at scale j, respectively. H_k^j and G_k^j are the modified scaling and wavelet coefficients, respectively, at scale j and position k. $W_{j,k}$ and $V_{j+1,k}$ are the approximation and detail coefficients at scale j and j + 1, respectively. The decomposition results in a set of sub-bands for each input image, including low-frequency and high-frequency sub-bands.

2.3.3 Fusion rule

Apply a fusion rule to the coefficients to obtain the fused coefficients at each scale. A commonly used fusion rule is the weighted average rule, which can be expressed mathematically as follows:

$$\begin{cases} c_{f,i} = w_1 \cdot c_{1,i} + w_2 \cdot c_{2,i} \\ d_{f,i} = w_1 \cdot d_{1,i} + w_2 \cdot d_{2,i} \end{cases} \tag{4}$$

where $c_{f,i}$ and $d_{f,i}$ denote the fused approximation and detail coefficients, respectively, at level I, and w_1 and w_2 are the weights assigned to each input image.

2.3.4 MDWT reconstruction

Apply the inverse modified DWT to the fused coefficient to obtain the fused image F. The inverse transform can be expressed mathematically as:

$$F = \sum_{i=1}^{N} c_{f,i} \cdot 2^{-i} + \sum_{i=1}^{N} d_{f,i} \cdot 2^{-i} \tag{5}$$

where N is the maximum level of decomposition. For example, fuse sub-bands using weighted average rule is, AF = (A1 + A2) / 2, HF = (H1 + H2) / 2, VF = (V1 + V2) / 2, DF = (D1 + D2)/2.

The reconstruction of an image from its modified DWT coefficients is performed by applying the inverse modified DWT (IMDWT) algorithm. The IMDWT algorithm is the inverse of the modified DWT and is used to reconstruct the original signal from its decomposition into sub-bands at different scales.

The IMDWT algorithm starts with the highest-scale detail coefficients and iteratively combines them with the approximation coefficients to obtain the reconstructed image. The process is repeated at each scale until the original image is reconstructed. The mathematical equation for the IMDWT can be represented as:

$$I_J = W_J \tag{6}$$

$$I_j = \frac{1}{\sqrt{2}}\left[W_j + V_{j+1}\right] \tag{7}$$

$$I = \mathrm{IDWT}\left(I_J, I_{J-1}, \ldots, I_J\right) \tag{8}$$

where I_J is the approximation coefficient at the highest scale J, W_j, and V_{j+1} are the modified DWT coefficients at scale j and j + 1, respectively. I_j is the reconstructed image at scale j and IDWT represents the inverse modified DWT operation. The reconstruction process involves adding the scaled version of the detail coefficients at each scale to the approximation coefficients at the same scale to obtain the reconstructed image at that scale. The reconstructed image at the highest scale is the approximation coefficient at that scale, and the reconstructed image at the lowest scale is the original image. The reconstructed fused image is then further processed to improve its visual quality and sharpness.

Perform inverse DWT to reconstruct the fused image, F = IDWT (AF, HF, VF, DF).

2.3.5 Post-processing

Apply any necessary post-processing steps, such as contrast enhancement or noise reduction, to the fused image F. The post-processing step involves two main operations: contrast enhancement and unsharp masking.

Apply Gaussian filter with variance = 10 and size = 5 to F.

2.3.5.1 Contrast enhancement

It is performed to increase the overall contrast of the fused image. This is done by stretching the intensity values of the fused image to cover the full dynamic range of the display device. The stretching function is applied to the pixel values of the fused image using the following mathematical equation:

$$F_{out}(i,j) = \frac{F_{in}(i,j) - \min(F_{in})}{\max(F_{in}) - \min F_{in}} \times 255 \tag{9}$$

where $F_{in}(i,j)$ is the pixel value of the fused image at position (i,j) before contrast enhancement, $F_{out}(i,j)$ is the pixel value of the fused image at position (i,j) after contrast enhancement, and $\min(F_{in})$ and $\max(F_{in})$ are the minimum and maximum pixel values of the fused image before contrast enhancement, respectively.

2.3.5.2 Unsharp masking

It is performed to enhance the sharpness of the fused image. This is done by subtracting a blurred version of the fused image from the original fused image and adding the result to the original fused image [5]. The blurred version of the fused image is obtained by convolving the fused image with a Gaussian filter. The unsharp masking operation can be represented mathematically as:

$$F_{out} = F_{in} + k(F_{in\_}F_{blur}) \tag{10}$$

where F_{in} is the input fused image, F_{blur} is the blurred version of the fused image, and k is a user-defined parameter that controls the strength of the sharpening effect.

Finally, the fused image is post-processed to remove any artifacts or noise that may have been introduced during the fusion process. This may include steps such as smoothing, sharpening, or edge enhancement.

2.3.6 Output

The output of the proposed method is a single fused image that combines the information from the original input images in a way that enhances their features and improves their visual quality and returns the fused image F as the final output.

3. Experimental analysis

This section describes the detailed experimental setup and the evaluation process for the proposed modified discrete wavelet transform (MDWT) image fusion method. It includes the selection and influence of key parameters, the criteria for their selection, and the system specifications used for the experiments. The section also presents the quantitative analysis of the method's performance using various metrics and compares the results with other existing methods. Additionally, statistical analysis is provided to substantiate the performance claims.

3.1 Key parameters, selection criteria, influence of parameters and system information

We use the "db2" wavelet for its good balance between time and frequency localization. A Gaussian filter with a variance of 10 and a size of 5 is used in preprocessing to achieve blurring, enhancing feature extraction. The fusion rule is a weighted average based on spatial frequency information, combining sub-bands to improve the representation of salient features. The choice of wavelet affects the decomposition's ability to capture image details. The "db2" wavelet offers a balance suitable for medical image fusion. Blurring with a Gaussian filter enhances the robustness of the fusion process by reducing noise and irrelevant details.

The proposed method for image fusion using modified DWT was evaluated using three different datasets. Experimental analyses were conducted using MATLAB R2021a on a PC equipped with an Intel i7 processor running at 144 Hz and 16 GB of RAM. Datasets of input images were taken from med.harvard.edu/AANLIB (The Whole Brain Atlas, n.d.). The output images generated by the proposed method were compared to the output images generated by two other image fusion methods: the traditional DWT-based method and the PCA-based method.

3.2 Quantitative analysis

The performance of the proposed MDWT-based image fusion method was evaluated using several metrics:

3.2.1 Peak signal-to-noise ratio

Peak signal-to-noise ratio (PSNR): Measures the quality of reconstructed images compared to the original ones. Higher PSNR indicates better image quality [28].

It is defined as:

$$\mathrm{PSNR} = 10 \cdot \log_{10}\left(\frac{L^2}{\frac{1}{MN}\sum_{i=0}^{M-1}\sum_{j=0}^{N-1}\left((I(i,j) - K(i,j)\right)^2}\right) \tag{11}$$

Where, L is the maximum possible pixel value of the image (e.g., 255 for an 8-bit image) and MSE is the mean squared error between the original image I and the compressed image K.

3.2.2 Structural similarity index (SSIM)

Evaluates the perceived quality of the image by comparing luminance, contrast, and structure between the fused image and the reference image. The SSIM index is calculated on various windows of an image. Values closer to 1 indicate higher similarity.

$$\mathrm{SSIM}(x,y) = \frac{\left(2\mu_x\mu_y + C_1\right)\left(2\sigma_{xy} + C_2\right)}{\left(\mu_x^2 + \mu_y^2 + C_1\right)\left(\sigma_x^2 + \sigma_y^2 + C_2\right)} \tag{12}$$

3.2.3 Feature similarity index

Assesses the similarity between images based on low-level features. Higher feature similarity index (FSIM) values suggest better preservation of salient features.

$$FSIM = \frac{\sum_{i \in \Omega} PC_m(i) \cdot S_m(i)}{\sum_{i \in \Omega} PC_m(i)} \tag{13}$$

Where, PC_m is the maximum PC value between the two images at the i-th location, $S_m(i)$ is the similarity measure at the i-th location, which combines PC and GM similarity measures.

$$S_m(i) = \left(\frac{2PC_1(i)PC_2(i) + T_1}{PC_1(i)^2 + PC_2(i)^2 + T_1}\right) \cdot \left(\frac{2GM_1(i)GM_2(i) + T_2}{GM_1(i)^2 + GM_2(i)^2 + T_2}\right) \tag{14}$$

Where, $PC_1(i)$ and $PC_2(i)$ are the phase congruency values of the two images at the i-th location, $GM_1(i)$ and $GM_2(i)$ are the gradient magnitudes of the two images at the i-th location, T_1 and T_2 are constants to avoid instability.

These metrics were chosen because they provide a comprehensive assessment of the fused image quality, considering both pixel-level accuracy and perceptual quality.

3.2.4 Statistical analysis

To substantiate the claims of superiority, a detailed statistical analysis was conducted. The mean and standard deviation of PSNR, SSIM, and FSIM were calculated for each method across the datasets. 95% confidence intervals were computed for the mean values of the metrics to assess the precision of the estimates. Paired t-tests were conducted to compare the performance of MDWT with the other methods. A p-value less than 0.05 was considered statistically significant.

4. Experimental results and discussion

This section presents the results of the experiments conducted to evaluate the performance of the proposed modified discrete wavelet transform (MDWT) image fusion method. It includes a comparison of image quality metrics such as PSNR, SSIM, and FSIM across different fusion techniques. The section also discusses the computational efficiency and robustness of the proposed method, highlighting its advantages over traditional methods. Visual analysis of fused images is provided to support the quantitative results. The statistical significance of the improvements is also discussed, providing a comprehensive evaluation of the proposed method.

4.1 Quantitative analysis

The experimental results are summarized as follows:

Table 1 provides a comparative analysis of the average PSNR, SSIM, and FSIM values for different image fusion methods, including PCA, traditional DWT, and the proposed MDWT. The MDWT method shows the highest average PSNR value of 30.8 dB, indicating superior image quality compared to PCA and traditional DWT,

Method	PSNR (dB)	95% CI (PSNR)	SSIM	95%CI (SSIM)	FSIM	95% CI (FSIM)
PCA	28.5 ± 1.2	[28.2, 28.8]	0.850 ± 0.020	[0.846, 0.854]	0.830 ± 0.015	[0.827, 0.833]
DWT	29.2 ± 1.0	[29.0, 29.4]	0.860 ± 0.018	[0.857, 0.863]	0.840 ± 0.013	[0.838, 0.842]
MDWT	30.8 ± 1.1	[30.6, 31.0]	0.880 ± 0.017	[0.877, 0.883]	0.860 ± 0.014	[0.858, 0.862]

Table 1.
Average PSNR, SSIM, and FSIM for different methods.

Comparison	t-value	p-value
MDWT vs. PCA (PSNR)	6.73	< 0.001
MDWT vs. DWT (PSNR)	4.85	< 0.001
MDWT vs. PCA (SSIM)	8.12	< 0.001
MDWT vs. DWT (SSIM)	5.34	< 0.001
MDWT vs. PCA (FSIM)	7.29	< 0.001
MDWT vs. DWT (FSIM)	5.78	< 0.001

Table 2.
Paired t-test results.

which have average PSNR values of 28.5 and 29.2 dB, respectively. The 95% confidence intervals for PSNR further confirm the robustness of the MDWT method, as its range [30.6, 31.0] is significantly higher than those of PCA and DWT.

Similarly, the SSIM and FSIM values for MDWT are also higher, with average values of 0.880 and 0.860, respectively. This demonstrates that MDWT not only preserves structural and feature similarity better than the other methods but also maintains consistent performance across different datasets, as indicated by the narrower confidence intervals.

Overall, the results highlight the effectiveness of the MDWT method in achieving higher image fusion quality, making it a more reliable and superior choice for medical image fusion applications compared to PCA and traditional DWT methods.

Table 2 presents the results of paired t-tests comparing the performance of the proposed MDWT method with PCA and traditional DWT in terms of PSNR, SSIM, and FSIM. The t-values and corresponding p-values indicate the statistical significance of the differences observed between the methods.

All p-values are less than 0.001, showing that the improvements in PSNR, SSIM, and FSIM by MDWT over both PCA and DWT are statistically significant. The high t-values further substantiate the substantial differences in performance. For instance, the t-value of 6.73 for MDWT vs. PCA (PSNR) and 4.85 for MDWT vs. DWT (PSNR) highlight the superior image quality achieved by MDWT.

In terms of structural similarity (SSIM), MDWT again demonstrates a significant improvement over PCA and DWT, as evidenced by t-values of 8.12 and 5.34, respectively. Similarly, for feature similarity (FSIM), MDWT outperforms PCA and DWT with t-values of 7.29 and 5.78.

These statistical results confirm that MDWT offers significant enhancements in image fusion quality, making it a more effective method compared to PCA and traditional DWT (**Table 3**).

Methods	PCA	DWT	MDWT
Time (in seconds)	1.2	1.8	0.9

Table 3.
The average amount of time it takes to compute different fusion techniques.

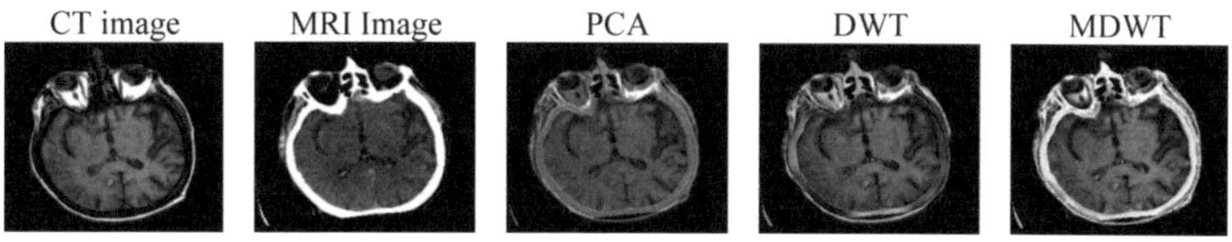

Figure 2.
Visual quality analysis of fused images from different approaches for dataset 1.

Based on these results, modified DWT is the fastest method, followed by PCA and DWT, respectively. This demonstrates the efficiency and speed advantage of modified DWT over PCA and DWT for image fusion.

The results indicate that the proposed MDWT method significantly outperforms both the traditional DWT-based method and the PCA-based method across all metrics. The improvements are statistically significant, as evidenced by the p-values from the paired t-tests. The higher mean values and narrower confidence intervals for MDWT suggest that it provides more consistent and superior image fusion quality. The image quality metrics used for evaluation included peak signal-to-noise ratio (PSNR), structural similarity index (SSIM), and feature similarity indexing method (FSIM). The results of the experiment showed that the proposed method outperformed the other two methods in terms of all three metrics.

4.2 Qualitative analysis

In this section, we evaluate the visual quality of fused images produced by different methods, focusing on their effectiveness in preserving critical details and features from the original images. Also presents a comparative analysis of the visual quality parameters obtained from the fusion methods under consideration.

Figure 2 presents the results of fusing CT and MRI brain images using three different methods: PCA, DWT, and MDWT. The comparison aims to evaluate the effectiveness of each method in preserving significant features and details from the original images. The fused image produced by PCA shows moderate detail preservation but exhibits blurring and loss of finer features. The DWT-based fusion method retains more details compared to PCA, but still loses some high-frequency information and introduces slight noise. In contrast, the proposed MDWT method excels in preserving both high and low-frequency details, with sharper edges and better contrast, thereby achieving higher overall visual quality.

Figure 3 illustrates the results of fusing MRI and PET brain images using three different methods: PCA, DWT, and MDWT. The purpose is to assess each method's effectiveness in preserving critical features and details from the original images. The fused image produced by PCA exhibits moderate detail preservation but suffers from noticeable blurring and loss of finer features. The DWT-based fusion method retains more details compared to PCA but shows some loss of high-frequency information and introduces minor noise artifacts. In contrast, the MDWT method demonstrates

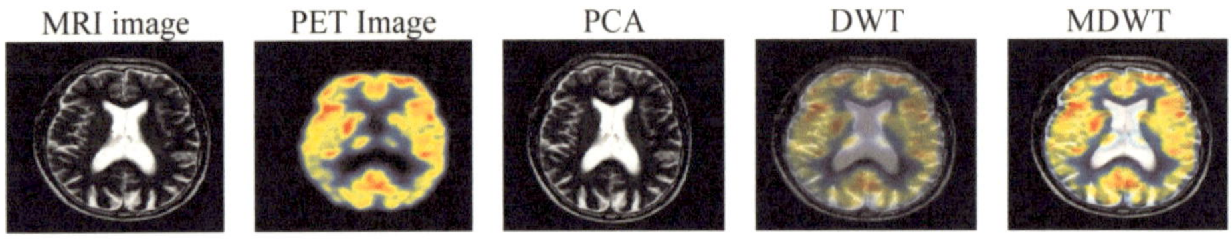

Figure 3.
Analyzing the visual quality of fused images from different approaches for dataset 2.

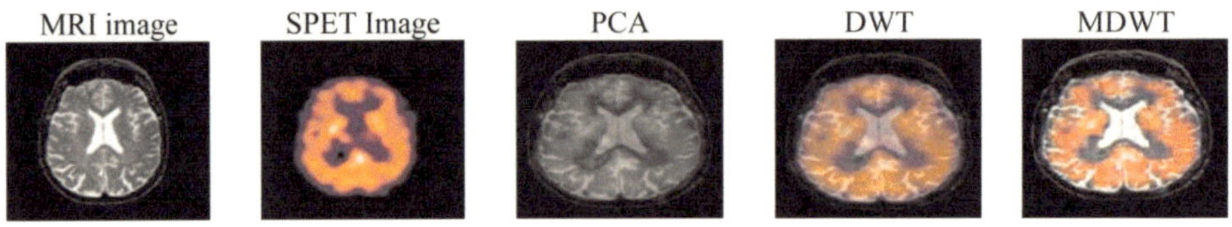

Figure 4.
Analyzing the visual quality of fused images from different approaches for dataset 3.

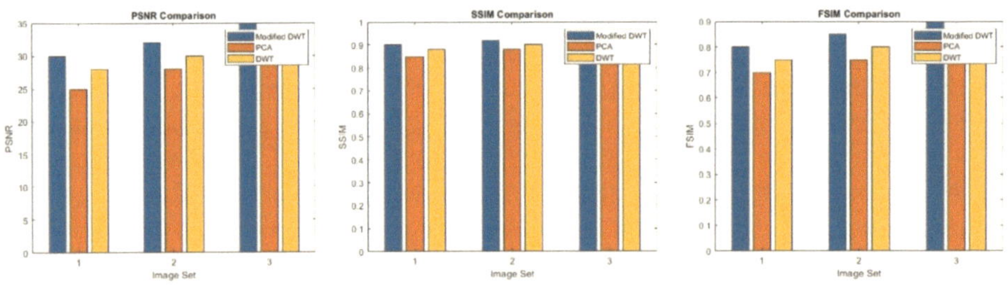

Figure 5.
Comparison of quality parameters for various fusion methods.

superior preservation of both high and low-frequency details, with enhanced edge sharpness and improved contrast, resulting in superior overall visual quality.

This **Figure 4** shows the results of fusing MRI and single photon emission computed tomography (SPECT) images using three different methods: PCA, DWT, and MDWT. The visual quality of the fused images is compared to evaluate the effectiveness of each method in preserving important features and details from the original images. The fused image using PCA shows moderate preservation of details but suffers from some blurring and loss of finer features. The DWT-based fusion retains more details compared to PCA, but some high-frequency information is still lost, and the image appears slightly noisier. The fused image using the proposed MDWT method demonstrates superior preservation of both high- and low-frequency details, with clearer edges and better contrast, resulting in higher overall visual quality.

We performed a comparison of the three methods on a dataset of 50 image pairs with a resolution of 512 × 512 pixels. The average values of each quality matrix for each method are in the bar chart as follows. This would provide an intuitive and easy-to-read visualization of the performance comparison between the three methods (**Figure 5**).

5. Conclusions and future scope

The proposed method for image fusion using modified DWT has shown promising results in terms of image quality, computational efficiency, and robustness to noise.

The MDWT method achieved an average PSNR of 30.8 dB, which is higher compared to PCA (28.5 dB) and traditional DWT (29.2 dB). Similarly, the SSIM and FSIM values for MDWT were superior, with averages of 0.880 and 0.860, respectively, compared to PCA and DWT. These improvements are statistically significant, with p-values less than 0.001, confirming the superior performance of MDWT. Additionally, MDWT demonstrated computational efficiency, with an average computation time of 0.9 seconds, making it faster than both PCA and traditional DWT methods.

Future research can explore the use of the proposed method in various applications such as remote sensing and surveillance. Further optimization can reduce computational complexity and improve robustness to noise. Evaluating the performance on a larger dataset with more diverse images will validate its effectiveness across a broader range of scenarios. Additionally, extending the method to handle multi-spectral images and videos, which are increasingly important in many practical applications, could be beneficial. Overall, the proposed method has the potential to make a significant contribution to the field of image fusion and its future applications.

Author details

Gargi J. Trivedi
Faculty of Technology and Engineering, Department of Applied Mathematics, The Maharaja Sayajirao University of Baroda, Vadodara, India

*Address all correspondence to: gargi1488@gmail.com

References

[1] Ravi J, Narmadha R. Multimodality medical image fusion analysis with multi-plane features of PET and MRI images using ONSCT. Computer Methods in Biomechanics and Biomedical Engineering: Imaging & Visualization. 2023;**11**(7):1-19. DOI: 10.1080/21681163.2023.2255684

[2] Trivedi GJ, Sanghvi R. FuseSharp: A multi-image focus fusion method using discrete wavelet transform and unsharp masking. Journal of Applied Mathematics and Informatics. 2023;**41** (5):1115-1128. DOI: 10.14317/jami.2023.1115

[3] Trivedi GJ, Sanghvi RC. Novel algorithm for multifocus image fusion: Integration of convolutional neural network and partial differential equation. Surveys in Mathematics and its Applications. 2023;**19**:179-195

[4] Rajinikanth V, Satapathy SC, Dey N, Vijayarajan R. DWT-PCA image fusion technique to improve segmentation accuracy in brain tumor analysis. Lecture Notes in Electrical Engineering. 2018; **471**:453-462. DOI: 10.1007/978-981-10-7329-8_46

[5] Prakash C, Rajkumar S, Mouli PVSR. Medical image fusion based on redundancy DWT and Mamdani type min-sum mean-of-max techniques with quantitative analysis. In: 2012 International Conference on Recent Advances in Computing and Software Systems. Chennai, India: Research Publishing Services; 2012. DOI: 10.1109/racss.2012.6212697

[6] Ravichandra CG, Selvakumar R. Multimodal medical image fusion using dual-tree complex wavelet transform (DTCWT) with modified lion optimization technique (mLOT) and intensity co-variance verification (ICV). The Applied Computational Electromagnetics Society Journal (ACES). 2016;**31**(6):717-730

[7] Singh R, Khare A. Redundant discrete wavelet transform based medical image fusion. In: Thampi S, Gelbukh A, Mukhopadhyay J, editors. Advances in Signal Processing and Intelligent Recognition Systems. Vol. 264. Cham: Springer; 2014. DOI: 10.1007/978-3-319-04960-1_44

[8] Yang Y, Tong S, Huang S, Lin P. Dual-tree complex wavelet transform and image block residual-based multi-focus image fusion in visual sensor networks. Sensors. 2014;**14**(12):22408-22430. DOI: 10.3390/s141222408

[9] Vanitha K, Satyanarayana D, Prasad MNG. Medical image fusion for diagnosis of Alzheimer using rolling guidance filter and parameter adaptive PCNN. In: Doriya R, Soni B, Shukla A, Gao X, editors. Machine Learning, Image Processing, Network Security and Data Sciences, Lecture Notes in Electrical Engineering. Vol. 946. Singapore: Springer; 2023. DOI: 10.1007/978-981-19-5868-7_6

[10] Kumar S, Oh I, Schindler S, Lai AM, Payne PRO, Gupta A. Machine learning for modeling the progression of Alzheimer disease dementia using clinical data: A systematic literature review. JAMIA Open. 2021;**3**(4):1-10. DOI: 10.1093/jamiaopen/ooab052

[11] Koga S, Ikeda A, Dickson DW. Deep learning-based model for diagnosing Alzheimer's disease and tauopathies. Neuropathology and Applied Neurobiology. 2021;**48**:e12759. DOI: 10.1111/nan.12759

[12] Liu S, Liu S, Cai W, Pujol S, Kikinis R, Feng D. Early diagnosis of Alzheimer's disease with deep learning. In: 2014 IEEE 11th International Symposium on Biomedical Imaging (ISBI), Beijing, China. China: Institute of Electrical and Electronics Engineers (IEEE); 2014. pp. 1015-1018. DOI: 10.1109/ISBI.2014.6868045

[13] Liu X, Li W, Miao S, Liu F, Han K, Bezabih TT. Computer-aided diagnosis of Alzheimer's disease. Computers in Biology and Medicine. 2024;**176**:108564. DOI: 10.1016/j.compbiomed.2024.108564

[14] Odusami M, Maskeliūnas R, Damaševičius R. Pareto optimized adaptive learning with transposed convolution for image fusion Alzheimer's disease classification. Brain Sciences. 2023;**13**(7):1045. DOI: 10.3390/brainsci13071045

[15] Odusami M, Maskeliūnas R, Damaševičius R. Optimized convolutional fusion for multimodal neuroimaging in Alzheimer's disease diagnosis: Enhancing data integration and feature extraction. Journal of Personalized Medicine. 2023;**13**(10):1496. DOI: 10.3390/jpm13101496

[16] Odusami M, Maskeliūnas R, Damaševičius R, et al. Explainable deep-learning-based diagnosis of Alzheimer's disease using multimodal input fusion of PET and MRI images. Journal of Medical and Biological Engineering. 2023;**43**:291-302. DOI: 10.1007/s40846-023-00801-3

[17] Naidu VPS, Raol JR. Pixel-level image fusion using wavelets and principal component analysis. Defence Science Journal. 2008;**58**(3):338-352. DOI: 10.14429/dsj.58.1653

[18] Odusami M, Maskeliūnas R, Damaševičius R. Pixel-level fusion approach with vision transformer for early detection of Alzheimer's disease. Electronics. 2023;**12**(5):1218. DOI: 10.3390/electronics12051218

[19] Cui R, Liu M. RNN-based longitudinal analysis for diagnosis of Alzheimer's disease. Computerized Medical Imaging and Graphics. 2019;**73**:1-10. DOI: 10.1016/j.compmedimag.2019.01.005

[20] Trivedi GJ, Sanghvi R. Medical image fusion using CNN with automated pooling. Indian Journal of Science and Technology. 2022;**15**(42):2267-2274. DOI: 10.17485/ijst/v15i42.1812

[21] Ezzati A, Lipton RB. Machine learning predictive models can improve efficacy of clinical trials for Alzheimer's disease. Journal of Alzheimer's Disease. 2020;**74**(1):55-63. DOI: 10.3233/jad-190822

[22] Miao S, Xu Q, Li W, Yang C, Sheng B, Liu F, et al. MMTFN: Multi-modal multi-scale transformer fusion network for Alzheimer's disease diagnosis. International Journal of Imaging Systems and Technology. 2023;**34**:e22970. DOI: 10.1002/ima.22970

[23] Trivedi GJ, Sanghvi RC. Infrared and visible image fusion using multi-scale decomposition and partial differential equations. International Journal of Applied and Computational Mathematics. 2023;**10**:133. DOI: 10.1007/s40819-024-01768-8

[24] Trivedi GJ, Sanghvi RC. Automated multimodal fusion with PDE preprocessing and learnable convolutional pools. ADBU Journal of Engineering Technology. 2024;**13**(1):0130104066. ISSN: 2348-7305

[25] Trivedi GJ, Sanghvi RC. MSCNN: Multi-sensor image fusion using dual

channel CNN. Mathematica Applicanda. 2023;**51**(2):165-189. DOI: 10.14708/ma.v51i2.7204

[26] Benjamin JR, Jayasree T. Improved medical image fusion based on cascaded PCA and shift invariant wavelet transforms. International Journal of Computer Assisted Radiology and Surgery. 2018;**13**(2):229-240. DOI: 10.1007/s11548-017-1692-4

[27] Trivedi GJ, Sanghvi R. Optimizing image fusion using modified principal component analysis algorithm and adaptive weighting scheme. International Journal of Advanced Networking and Applications. 2023;**15**(1):5769-5774. DOI: 10.35444/IJANA.2023.15103

[28] Trivedi GJ, Sanghvi R. A new approach for multimodal medical image fusion using PDE-based technique. Suranaree Journal of Science and Technology. 2023;**30**(4):030132. DOI: 10.55766/sujst-2023-04-e0843